Mahmood Alam, Abhishek Dwivedi, Anshuman Srivastava, Ajay Kumar Mishra
Manufacturing Processes

Also of interest

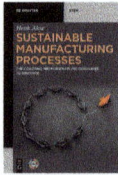

Sustainable Manufacturing Processes.
The Coaching Method Enabling Companies to Innovate
Henk Akse, 2025
ISBN 978-3-11-138342-2, e-ISBN (PDF) 978-3-11-138366-8,
e-ISBN (EPUB) 978-3-11-138380-4

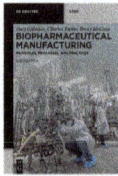

Biopharmaceutical Manufacturing.
Principles, Processes, and Practices
2nd, Completely Revised and Extended Edition
Gary Gilleskie, Charles Rutter, Becky McCuen, 2025
ISBN 978-3-11-111206-0, e-ISBN (PDF) 978-3-11-111245-9,
e-ISBN (EPUB) 978-3-11-111396-8

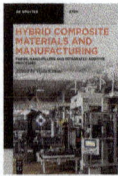

Hybrid Composite Materials and Manufacturing.
Fibers, Nano-Fillers and Integrated Additive Processes
Vipin Kumar (Ed.), 2024
ISBN 978-3-11-101934-5, e-ISBN (PDF) 978-3-11-101954-3,
e-ISBN (EPUB) 978-3-11-102013-6

Advanced Process Engineering Control
2nd, Revised and Extended Edition
Paul Serban Agachi, Mircea Vasile Cristea, Alexandra Ana Csavdari,
Botond Szilagyi, 2023
ISBN 978-3-11-078972-0, e-ISBN (PDF) 978-3-11-078973-7,
e-ISBN (EPUB) 978-3-11-078986-7

Mahmood Alam, Abhishek Dwivedi,
Anshuman Srivastava, Ajay Kumar Mishra

Manufacturing Processes

Metal Forming, Machining, Metal Casting, Welding
and Additive Manufacturing

DE GRUYTER

Authors
Er. Mahmood Alam
Department of Mechanical Engineering
Integral University
226026 Lucknow
Uttar Pradesh
India
mahmood@iul.ac.in

Dr. Abhishek Dwivedi
Department of Mechanical Engineering
Integral University
226026 Lucknow
Uttar Pradesh
India
abhishek@iul.ac.in

Prof. Dr. Anshuman Srivastava
Indian Institute of Packaging (IIP)
Sarojini Nagar
226008 Lucknow
Uttar Pradesh
India
anshuman0203@gmail.com

Prof. Dr. Ajay Kumar Mishra
Department of Chemistry
University of the Western Cape
Robert Sobukwe Rd
7535 Cape Town, Bellville
Republic of South Africa (RSA)
ajaykmishraedu@gmail.com

ISBN 978-3-11-914753-8
e-ISBN (PDF) 978-3-11-220589-1
e-ISBN (EPUB) 978-3-11-220628-7

Library of Congress Control Number: 2025935647

Bibliographic information published by the Deutsche Nationalbibliothek
The Deutsche Nationalbibliothek lists this publication in the Deutsche Nationalbibliografie;
detailed bibliographic data are available on the Internet at http://dnb.dnb.de.

© 2025 Walter de Gruyter GmbH, Berlin/Boston, Genthiner Straße 13, 10785 Berlin
Cover image: Byjeng/iStock/Getty Images Plus
Typesetting: Integra Software Services Pvt. Ltd.

www.degruyterbrill.com
Questions about General Product Safety Regulation:
productsafety@degruyterbrill.com

Preface

Manufacturing plays a crucial role in today's industrial world by transforming raw materials into products that meet human needs, and a solid understanding of its principles and practices is essential for every aspiring engineer and technocrat. This book is crafted to provide a balanced blend of theoretical concepts and practical applications, catering primarily to first-year engineering students across all disciplines, while also serving as a refresher for second- and third-year students. It covers a wide range of manufacturing processes—from traditional methods like casting, forming, and machining to advanced techniques such as advanced and additive manufacturing—offering readers a comprehensive foundation in the subject. Written in clear and simple language, the content aligns with the syllabi of various universities and technical boards, making it suitable for both diploma and degree-level courses. To support effective learning, the book features practice questions, comparison tables, and well-illustrated diagrams that bring clarity to tools, machines, and manufacturing setups. This book is intended to be a useful guide for the new learners or the readers brushing up their knowledge. While every effort has been made to ensure accuracy, the author welcomes constructive feedback and suggestions to improve future editions. It is hoped that this book will serve not only as a valuable academic resource for students but also as a practical reference for professionals working in the manufacturing industry.

https://doi.org/10.1515/9783112205891-202

Contents

About the book

Manufacturing has always been an important part of the industrial environment, enabling the production of products for the service of mankind. The knowledge of manufacturing practices is highly essential for all engineers and technocrats to familiarize themselves with modern concepts of manufacturing technologies. The primary purpose of this manuscript is to provide theoretical and practical knowledge of manufacturing processes to all first-year engineering students across all branches, as well as second- and third-year students, to help them recall the basics and deepen their understanding of manufacturing science. This book offers a concise overview of essential manufacturing processes, providing readers with a foundational understanding of the methods used to transform raw materials into finished products. It covers traditional techniques such as casting, forming, and machining, alongside modern methods like additive manufacturing. By exploring both the theoretical and practical aspects of these processes, the book serves as a valuable resource for students, engineers, and professionals in the manufacturing industry. Whether you are new to the field or seeking to refresh your knowledge, this guide will equip you with the insights needed to navigate the world of manufacturing.

This book covers most of the syllabus of manufacturing processes and technology for engineering diploma and degree classes as prescribed by different universities and state technical boards. While preparing the manuscript of this book, the examination requirements of engineering students have also been kept in mind by including practice test questions. The book is written in very simple language so that even an average student can easily grasp the subject matter. Some comparisons are provided in tabular form, and emphasis has been placed on figures to enhance understanding of tools, equipment, machines, and manufacturing setups, ranging from traditional to advanced manufacturing, as used in various manufacturing shops.

The whole text has been organized into seven chapters.

Chapter 1 presents a brief introduction to materials and alloys, including the fundamentals of ferrous materials, non-ferrous materials, melting furnaces, properties and testing of engineering materials, and heat treatment.

Chapter 2 provides the necessary details of the mechanical working of metals. Fundamental concepts related to forging work and other mechanical working processes (hot and cold working) are discussed at length with clear and detailed sketches.

Chapter 3 provides construction and operational details of various conventional machine tools, namely lathe, drilling machine, shaper, planer, slotter, and milling machine, including some unconventional machines like EDM, ECM, AJM, etc., with the help of neat diagrams.

Chapter 4 describes the manufacturing process of casting, including pattern making and component production using molds.

https://doi.org/10.1515/9783112205891-204

Chapter 5 provides necessary details about various welding and allied joining processes, such as gas welding, arc welding, resistance welding, solid-state welding, brazing, and soldering, along with their applications and defects.

Chapter 6 covers the basics of additive manufacturing and provides a brief overview of the latest methods involved in it, such as 3D printing, SLS, FDM, etc.

Chapter 7, the last chapter of the book, discusses the basic concepts of the economic importance of manufacturing, plant location, and plant layout with its types, production vs. productivity, common types and uses of wood, cement concrete, ceramics, rubber or elastomers, composite materials, powder metallurgy processes and applications, methods of plastic product manufacturing, galvanizing and electroplating, modern trends in manufacturing, automation including CAD/CAM, CIM, FMS, and general industrial safety measures to be followed in various manufacturing shops, all described in detail.

The author strongly believes that the book will serve not only as a text book for students of the engineering curriculum but also as a reference material for engineers working in manufacturing industries. Although every care has been taken to check for misprints and mistakes, it is difficult to claim perfection. Any errors, omissions, or suggestions for improvement of this volume will be gratefully acknowledged and included in the next edition.

Chapter 1
Basic metals and alloys: properties and applications

1.1 Introduction: mechanical properties of materials

The characteristics of a material that are linked to its resistance to loads and mechanical forces are known as its mechanical properties.

OR

A material's mechanical properties are the characteristics that define how it reacts to loads and applied forces. When determining whether a material is appropriate for a given application, these qualities are crucial.

1.1.1 Basic properties of materials

i. **Strength:** This refers to a material's ability to endure external forces without breaking:
    a) Compressive strength,
    b) Tensile strength,
    c) Bending strength,
    d) Torsional strength, and
    e) Shear strength.
ii. **Stiffness:** It is a material's capacity to withstand deformation or deflection in material under external load. This property makes the material rigid and resistant to bending.
iii. **Ductility:** When a tensile force is applied, the material's property allows it to be pulled into wire. Typically, ductility is expressed as a percentage of length elongation and cross-sectional area decrease:

$$\% \text{ Elongation in length} = \frac{\text{Change in length}}{\text{Original length}} \times 100$$

$$\% \text{ Reduction in cross-sectional area (C/S)} = \frac{\text{Change in C/S area}}{\text{Original C/S area}} \times 100$$

iv. **Toughness:** It is a material's ability to withstand breaking under heavy impact stresses such as a hammer strike. It is determined by measuring the energy absorbed by a unit volume of the material following tension up to the fracture point. This characteristic is ideal for components that are subjected to impact and shock loads. The basic view of toughness for ductile and brittle materials has been discussed in Figure 1.1.

https://doi.org/10.1515/9783112205891-001

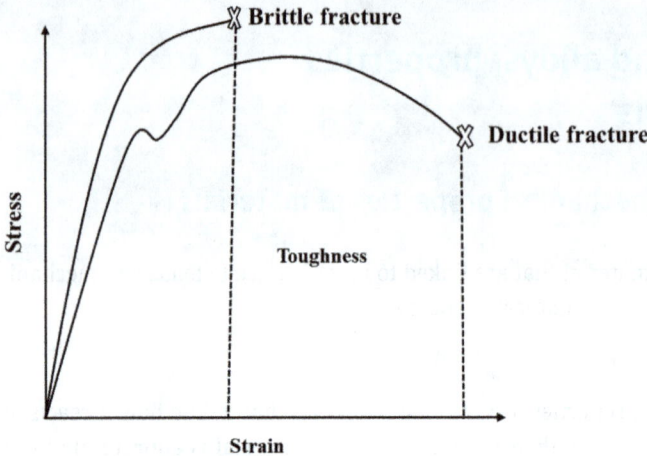

Figure 1.1: Stress-strain graphs showing the toughness of brittle and ductile materials.

v. **Elasticity:** When an external force is removed from a material that has deformed, it has the ability to return to its previous shape. Materials for machinery and tools should have this quality.

vi. **Plasticity:** Permanent deformation is the ability of a substance to maintain its original shape after being compressed or stretched. This is crucial for producing items like jewelry, stamped coins, and forged parts.

vii. **Malleability:** The ability to roll or smash the material into thin sheets is a unique instance of ductility. Although it is not necessary for a malleable substance to be strong enough to be pulled into wires, it should be plastic. For example, copper, aluminum, wrought iron, soft steel, and lead.

viii. **Brittleness:** This feature of a substance is the antithesis of ductility. It is the property of a material that allows for abrupt breaking with minimal long-term bending or stretching. Brittle materials do not exhibit significant elongation or deformation when tugged or stretched; instead, they break apart.

ix. **Hardness:** It has many different meanings and is one of the metals' most significant characteristics. It encompasses a wide range of qualities, including machinability, deformation resistance, and resistance to wear and scratches. It gives the metal the capacity to cut another metal.

1.1.2 Fracture, fatigue, and creep

When an object experiences active forces well below its melting point, it may split into two or more parts is called fracture.

There are two steps in the fracture process:
i.   Crack initiation
ii. Propagation of crack

The different types of fracture are as follows:
i.   Ductile fracture
ii. Brittle fracture
iii. Fatigue fracture
iv. Creep fracture

Ductile fracture can further be classified and represented in Figure 1.2:
a) Highly ductile fracture;
b) Necking and a somewhat ductile fracture the most prevalent type of ductile fracture, known as a cup-and-cone fracture.

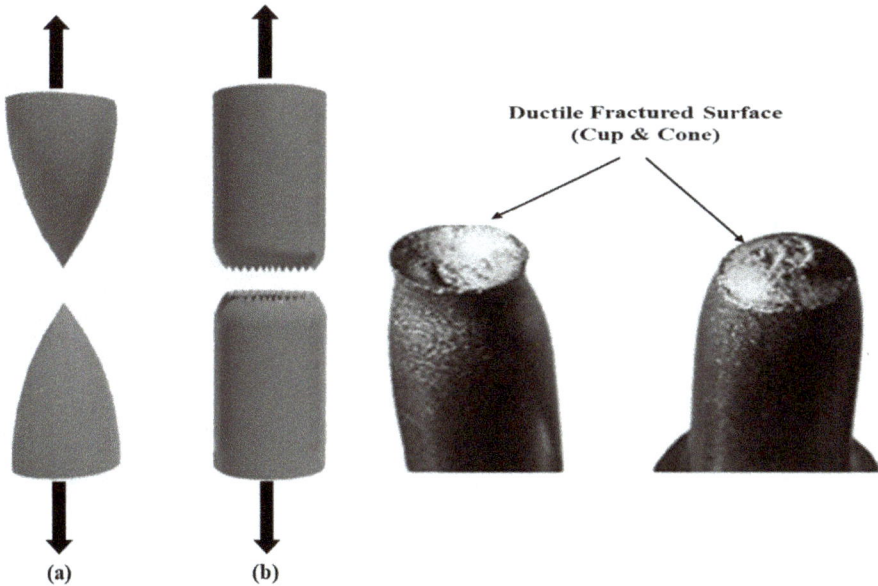

Ductile Fractured Surface (Cup & Cone)

(a)   (b)

Figure 1.2: Ductile fracture: (a) highly ductile; (b) moderate ductile (cup-and-cone).

1.1.3 Ductile fracture

Ductile fracture is characterized by the fracture or failure of a material accompanied by significant deformation. This type of fracture occurs through the slow propagation of a crack after extensive and noticeable deformation. The crack typically propagates at a 45° angle, where the shear stress is at its maximum, leading to a cup-and-cone appearance on the fracture surface. The stages depicted in Figure 1.3 involved in ductile fracture are as follows:

(a) Initial necking
(b) Cavity formation
(c) Cavities going to form a crack
(d) Crack propagation
(e) Final shear/fracture/separation of material

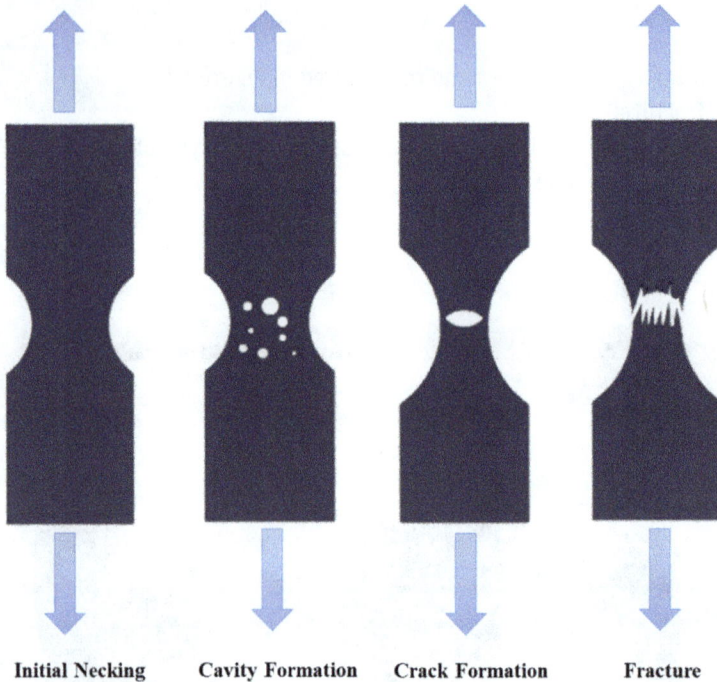

| Initial Necking | Cavity Formation | Crack Formation | Fracture |

Figure 1.3: Stages in ductile fracture (cup and cone).

1.1.4 Brittle fracture

Brittle fracture is defined as the separations of metal without any appreciate deformation. Metal get fracture with little or no plastic deformation (no necking as shown in Figure 1.4) and material absorb minimum amount of energy before failure (low toughness). This type of fracture occurs in brittle material. Crack propagation is rapid and propagates perpendicular to the direction of the applied stress.

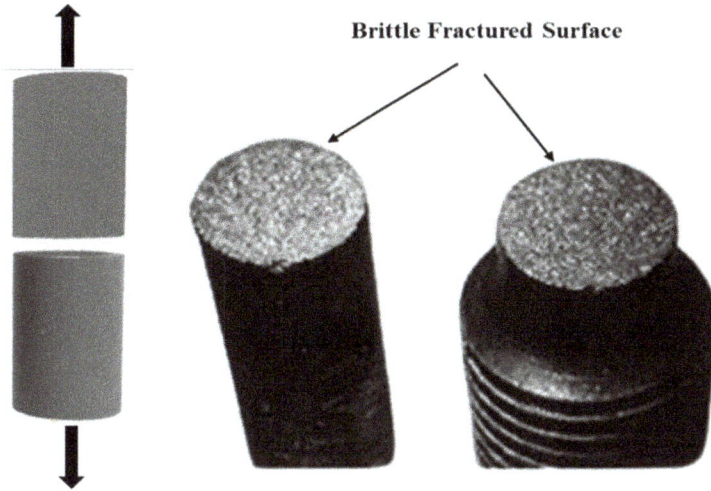

Brittle Fractured Surface

Figure 1.4: Brittle fracture.

1.1.5 Fatigue fracture (failure)

When a material is exposed to fluctuating and reversing loads as shown in Figure 1.5, it can fail due to fatigue. These changing loads cause tiny cracks to form and grow, eventually leading to failure. This type of failure happens at stress levels much lower than the material's tensile strength and within its elastic range. The risk of fatigue failure increases with higher temperatures and faster rates of straining. Fatigue failure is especially important to consider in parts like rotating shafts, axles, and airplane wings.

1.1.6 Creep fracture

Creep is a solid material's propensity to move slowly or permanently deform when stressed in various stages as shown in Figure 1.6. Creep can also be explained as a slow permanent deformation of a metal due to prolonged steady loading. It is influenced by temperature so the material for long period. Some material like copper iron and its alloys are susceptible to creep when subjected to high temperature. Some materials like lead, zinc, and rubber also creep at room temperature. Gradual loosening of bolts and sagging of long span cables are the examples of creep failure. Figure 1.7 represents example of creep failure in long span transmission cables.

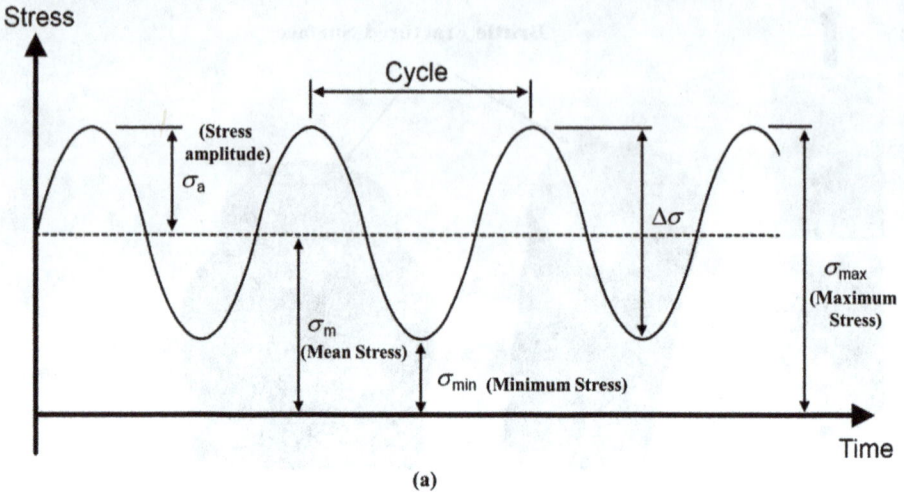

(a)

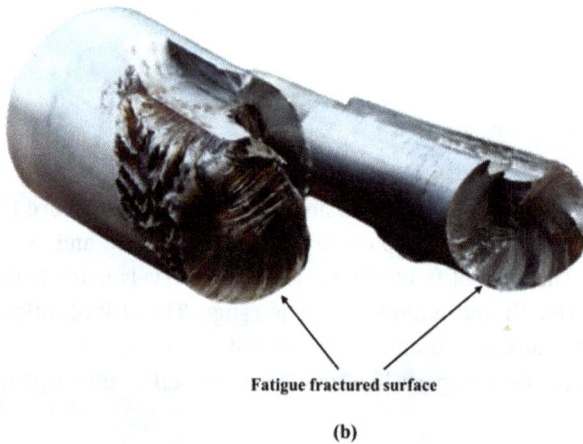

Fatigue fractured surface

(b)

Figure 1.5: (a) Reversible and fluctuating stress causing fatigue fracture with time; (b) example of fatigue fracture.

Stages in creep

i. Primary creep (transient creep): This initial stage features a relatively high rate of deformation that decreases over time. The material undergoes significant strain, but the rate of strain (deformation per unit time) slows down as the material hardens and becomes more resistant to further deformation (Figure 1.6).

ii. Secondary creep (steady-state creep): The deformation rate becomes constant, reaching a balance between strain hardening and recovery processes. Also known as steady-state creep, this stage is characterized by a constant rate of deformation. The material deforms at a steady, linear rate, which is considered the most predictable and stable part of the creep process. The balance between work hardening and recovery processes is achieved during this stage.

iii. Tertiary creep: The deformation rate accelerates rapidly leading to the formation of cracks, voids, and eventual failure of the material.

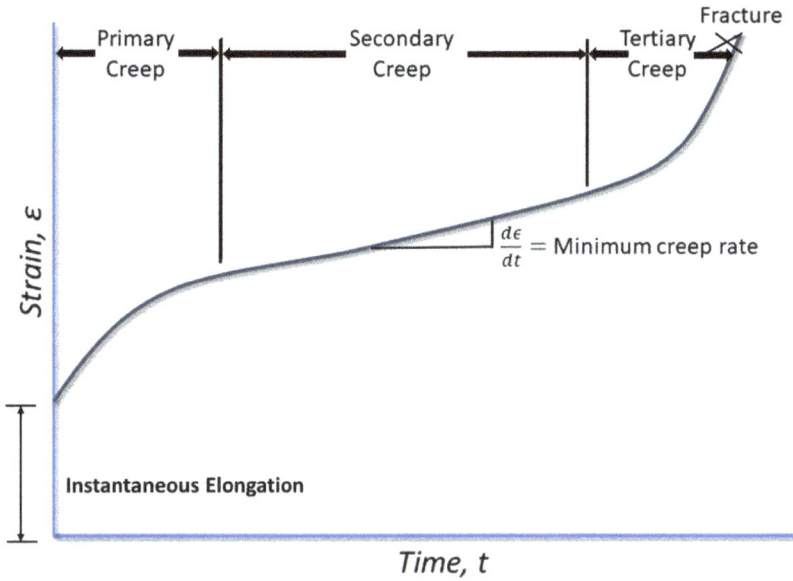

Figure 1.6: Stages in creep failure.

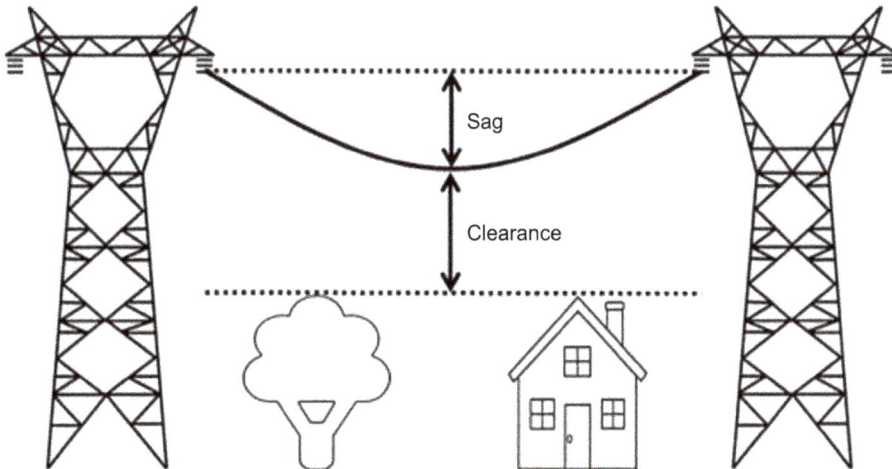

Figure 1.7: Example of creep failure.

Microstructural changes such as void formation, grain boundary sliding, and crack development contribute to the material's increasing instability and eventual rupture. This stage often culminates in the material fracturing or breaking apart.

Understanding these stages helps in predicting the lifespan of materials under high-temperature conditions and is crucial in designing components for applications like turbines, boilers, and engines where creep is a significant concern.

1.2 Ferrous material

Figure 1.8 provides a generic classification of engineering materials. Engineering materials fall into two general categories: metallic and non-metallic. Ferrous and non-ferrous materials are further classifications for metallic materials. The paragraph that follows will now address ferrous materials.

Materials classified as ferrous have iron as their primary ingredient. Additional components that come in different amounts to create different ferrous compounds include carbon, manganese, silicon, sulfur, and phosphorus.

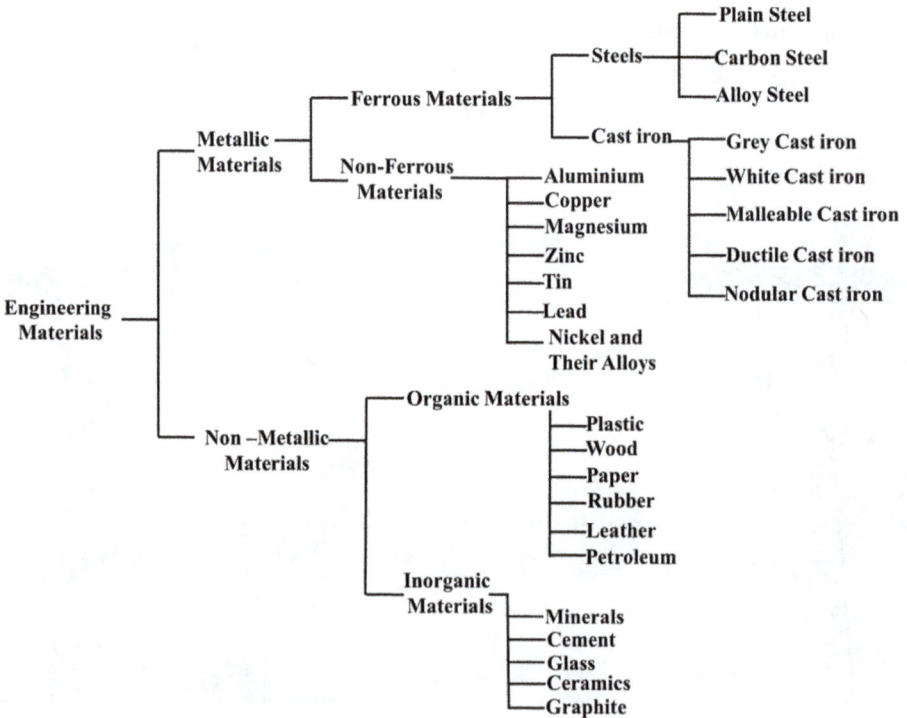

Figure 1.8: Material classification.

The materials listed below are most frequently utilized in engineering applications:
a) Wrought iron
b) Cast iron
c) Carbon steel
d) Stainless steel
e) Tool steel

1.2.1 Wrought iron

Wrought iron is a tough, malleable iron alloy with less than 0.08% carbon, making it easily shaped and welded. It has a grainy texture and resists corrosion better than pure iron. Historically, it was used in construction, tools, and decorative items. Today, it is mainly used for ornamental purposes and restoration work. Regular painting and cleaning help maintain its appearance and prevent rust.

Properties:
i.   Ductility and malleability: Wrought iron can be easily shaped and formed.
ii. Toughness: It is resilient and not prone to cracking or breaking under stress.
iii. Corrosion resistance: It resists corrosion better than pure iron, making it suitable for outdoor use.
iv. Weldability: It can be welded easily due to its low carbon content.
v.   Aesthetic appeal: It has a grain resembling wood, which adds to its visual appeal in decorative applications.

Modern uses: Wrought iron is still utilized as ornamental elements and in restoration projects to maintain historical authenticity despite mild steel having largely supplanted it due to cost and availability.

Historical uses:
a) Construction: Structural components such as beams, columns, and bridges
b) Decorative ironwork: Gates, railings, and fencing
c) Tools and hardware: Chains, hooks, nails, and other hardware
d) Art and craft: Decorative items and sculptures

1.2.2 Cast iron with properties and applications

Cast iron is an iron-carbon alloy that contains between 2% and 4% carbon. It is renowned for having outstanding strength, wear resistance, and castability.

Properties:
a) High carbon content: 2–4%, making it hard and brittle
b) Excellent castability: Easily molded into complex shapes
c) Good compressive strength: Strong under compressive loads
d) Wear resistance: Durable and resistant to wear
e) Poor ductility: Brittle and can fracture under tensile stress
f) Good fluidity: Flows easily when molten, filling molds completely

Types of cast iron:
i. Gray cast iron
ii. Malleable cast iron
iii. Nodular cast iron
iv. White cast iron
v. Mottled cast iron
vi. Alloy cast iron

1. Gray cast iron: Gray cast iron gets its name from the gray appearance of its shattered sections. As shown in Figure 1.9, it contains graphite flakes, which are free of carbon. Because of the concentration of stress at the edges of these flakes, the material exhibits brittle behavior and a very low tensile strength. Nonetheless, it has good compressive strength. Table 1.1 presents the composition, key characteristics, and common applications of gray cast iron, providing a foundational understanding of its material properties and industrial relevance.

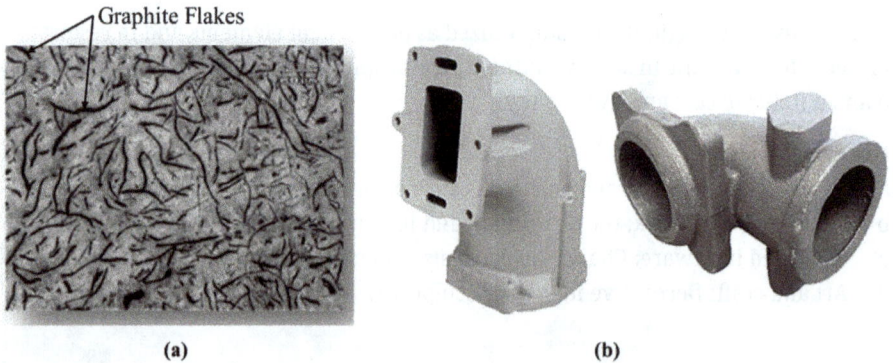

Graphite Flakes

(a)   (b)

Figure 1.9: (a) Microstructure of gray cast iron; (b) product made by gray cast iron.

2. White cast iron: In white cast iron carbides (Fe_3C) are present which is white in color. The fractured surface therefore appears white. It is only white cast iron in which carbon present in form of carbides. Due to iron-carbide (Fe_3C) white cast iron become extremely hard. Table 1.2 outlines the composition, distinct properties, and typical uses

Table 1.1: Composition, characteristics, and applications of gray cast iron.

Composition	Characteristics	Applications
– Carbon: 3–3.5% – Silicon: 1–2.75% – Phosphorous: 0.15–1% – Sulfur: 0.02–0.15% – Manganese: 0.4–1% – Rest % is of iron	– Good machinability – Good compressive strength and low tensile strength. – Free graphite present in the matrix impart lubricating properties – Good resistance to wear – Good damping capacity – It is brittle in nature	– Pipe fittings – Manhole covers – Piston, flywheel – Structures subjected to Compressive load – Machine tool structures

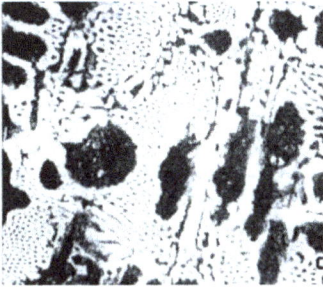  

(a)   **(b)**

Figure 1.10: (a) Microstructure of white cast iron; (b) product made by white cast iron.

Table 1.2: Composition, characteristics, and applications of white cast iron.

Composition	Properties	Applications
– Carbon: 1.75–2.3% – Silicon: 0.85–1.25% – Phosphorous: 0.05–0.2%. – Sulfur: up to 0.12% – Manganese: up to 0.5% – Rest % is of iron	– The presence of carbide (Fe_3C) makes it extremely hard and brittle. – It has poor machinability. – It has high tensile strength but low compressive strength. – Its melting point is higher than that of gray cast iron. – It offers excellent resistance to abrasion.	– It is essential for manufacturing parts that experience excessive wear or require a hard outer surface, such as wheels, cams, balls, and brake shoes. It is also utilized as a raw material to make wrought iron and malleable cast iron.

of white cast iron, offering insight into its behavior and suitability for various industrial applications. Microstructure and few applications are shown in Figure 1.10.

3. Malleable cast iron: White cast iron (cementite) is created when cast iron is quickly cooled because graphite flakes do not form. After that, this white cast iron is heated to about 930 °C and maintained there for a long time in an iron oxide atmosphere. Cementite (Fe_3C) breaks down into ferrite and free graphite at this temperature. The combined carbon further breaks down into graphite when it cools. More free carbon is released if the cooling process proceeds very slowly. Malleablizing is the term used to describe this release of free carbon. Table 1.3 highlights the composition, unique characteristics, and practical applications of malleable cast iron, helping to understand its adaptability and performance in different engineering contexts.

Table 1.3: Composition, characteristics, and applications of malleable cast iron.

Composition	Properties	Applications
– Carbon: 2–3% – Silicon: 0.6–1.3% – Phosphorous: up to 0.18% – Sulfur: 0.02–0.20% – Manganese: 0.2–0.6% – Rest % is of iron	– Excellent machinability – Good ductility, can be bent without breaking – Shock-resistant – High yield strength – Good wear resistant	– It is incredibly hard and brittle because of the presence of carbide (Fe_3C). – It is not very machinable. – Its compressive strength is low, but its tensile strength is high. – Compared to gray cast iron, it has a greater melting point. – It provides outstanding abrasion resistance.

4. Nodular cast iron: Nodular cast iron is distinguished from other types of cast iron by having better ductility and toughness. It is sometimes referred to as spheroidal graphite iron or ductile iron. Better mechanical qualities are imparted by the graphite in nodular cast iron, which is present in the form of spheroids or nodules rather than flakes can be visualize in Figure 1.11.

In this cast iron spherical carbon nodules structure is developed. If magnesium, cerium, etc. are added to the melt with very low sulfur content (less than 0.05%), this will inhibit carbon from forming into flakes. The carbon forms into nodules or spheres in the matrix of ferrite or pearlite. Table 1.4 provides an overview of the composition, key properties, and common applications of nodular cast iron, emphasizing its strength, ductility, and versatility in engineering uses.

1.2.3 Carbon steels

Steel is an iron and carbon alloy that contains up to 1.5% carbon in the form of iron carbide. General classification is given in Figure 1.12. A summary of carbon steels and alloy steels is presented in Table 1.5, highlighting their compositions, mechanical

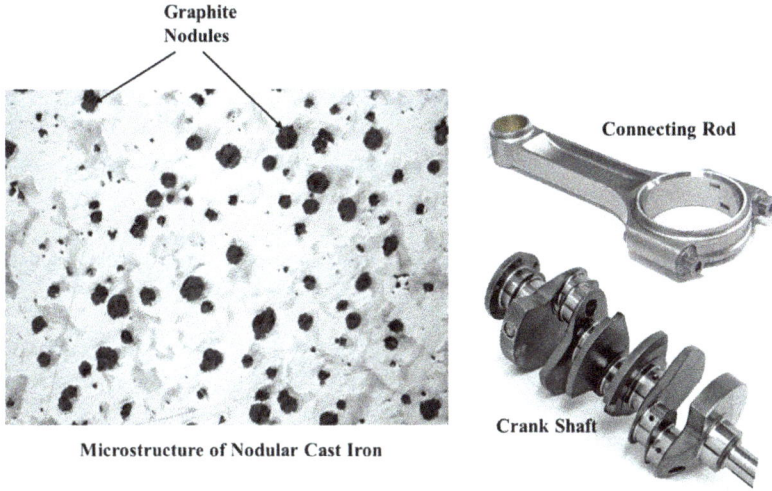

Figure 1.11: Microstructure of nodular cast iron and product made by the nodular cast iron.

Table 1.4: Composition, characteristics, and applications of nodular cast iron.

Composition	Properties	Applications
– Carbon: 3.2–4.2% – Silicon: 1–2.8% – Manganese: 0.3–0.8% – Phosphorus: 0.03–0.1% – Magnesium: 0.05–0.1% – Rest is iron.	– Higher ductility – Superior mechanical properties and wear resistance. – Good machinability and excellent castability due to good flowability – Damping capacity between cast iron and steel	– Crankshafts, truck axles, wheel hubs, engine connecting rods, and front wheel spindle supports are among the automotive parts that employ it. Power gearbox yokes, high-temperature applications such as turbo housings and manifolds, and high-security valves for a variety of uses are also among its uses.

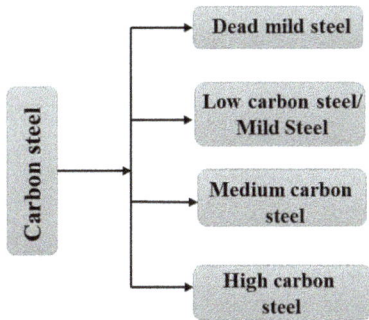

Figure 1.12: Classification of carbon steel.

Table 1.5: Summary of carbon steels and alloy steels.

Carbon steel	Properties	Applications
Dead mild steel (up to 0.15% carbon content)	1. Excellent ductility 2. Excellent forming characteristics 3. Higher toughness 4. Excellent machinability	Wires, thin sheets, rods, etc. with great ductility
Low carbon steel (C = 0.15–0.32%) also known as mild steel	1. Soft and ductile 2. Toughness is good but has low wear resistance 3. Good forming characteristics 4. It is challenging to harden using the heat treatment method 5. Carburizing is possible in certain unique circumstances	Steel wires, chains, rivets, nuts, bolts, thin canes, and tiny forgings, among other items
Medium carbon (C = 0.35–0.5%)	1. Less ductile than mild steel yet harder 2. More heat-treatment responsiveness than mild steel 3. A strong tensile strength of 600–710 N/mm^2 4. Can be easily welded or brazed	Structural steel section, rails and garden tools, connecting rods, gear, crankshafts axles, and drop forgings
High carbon steel (C = 0.55–1.5%) also known as tool steel	1. The simple carbon steel family's highest tensile strength and hardness 2. A variety of heat treatment techniques make it simple to regulate mechanical qualities 3. Plain carbon steel has the lowest ductility and machinability	Tools, drill punches, saw blades, chisels, hand tools, etc.

properties, and typical applications to illustrate their roles in various industrial and structural settings.

Table 1.6: Properties and applications of alloy steel.

Alloy steel	Properties	Applications
Stainless steel with 18% chromium, 8% nickel, and 8% magnesium	Sturdy, resilient, and resistant to corrosion	For creating surgical tools, kitchenware, and sinks
Carbon steel alloyed with tungsten, chromium, and vanadium is known as tool steel	– They possess high hardness – Even at high temperatures, resistant to frictional wear – Only ground; difficult to machine	Drills and metal cutting instruments (lathe and milling)

Properties and applications of alloy steel including stainless steel and carbon steel is compared in Table 1.6.

1.3 Heat treatment

1.3.1 Elementary introduction

This process alters the attributes and features of metals through controlled heating and cooling. Cold working induces stress in metals, leading to work hardening, which hinders further working. To soften the metal and allow additional processing, it can undergo heat treatment. A complete cycle of heat treatment is shown in Figure 1.13.

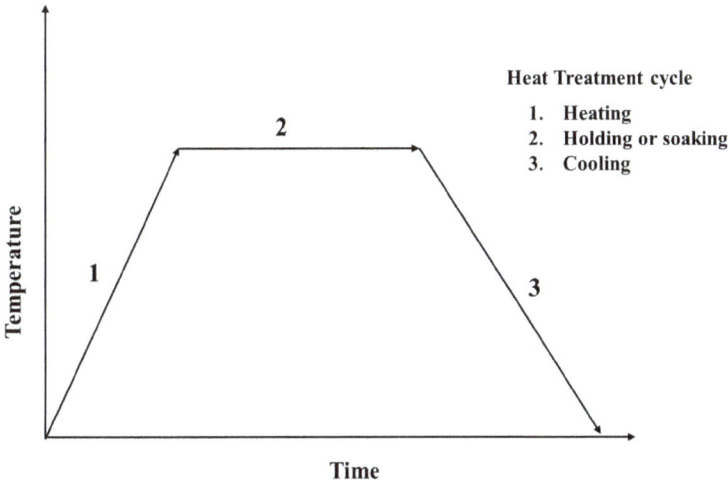

Figure 1.13: Stages of heat treatment process.

Heat treatment is done in three steps:
i.   Bring the work piece's temperature up to the desired level.
ii. Depending on the thickness of the piece, keep it at that temperature for the necessary amount of time (soaking).
iii. Cool it in a controlled manner to impart to it the desired properties

1.3.2 Purpose of heat treatment

–   Improve hardness: To increase the hardness of the metal for better wear resistance.
–   Enhance ductility: To improve ductility, making the metal more malleable and easier to work with.

– Relieve stress: To reduce internal stresses brought on by casting, machining, welding, and other procedures.
– Refine grain structure: To improve mechanical qualities and performance by fine-tuning the grain structure.
– Improve toughness: To increase toughness and impact resistance.
– Modify electrical and magnetic properties: To tailor electrical conductivity and magnetic properties for specific applications.
– Improve machinability: To make the metal easier to machine by softening it or improving its cutting characteristics.
– Increase strength: To enhance the strength of the metal, making it more durable for structural applications.
– Enhance wear resistance: To improve wear resistance, extending the life of the component in abrasive environments.
– Prepare for further processing: To prepare the metal for subsequent manufacturing processes such as forming, welding, or machining.

1.3.3 Classification of heat treatment process

Figure 1.14 illustrates how the heat treatment procedures might be categorized.

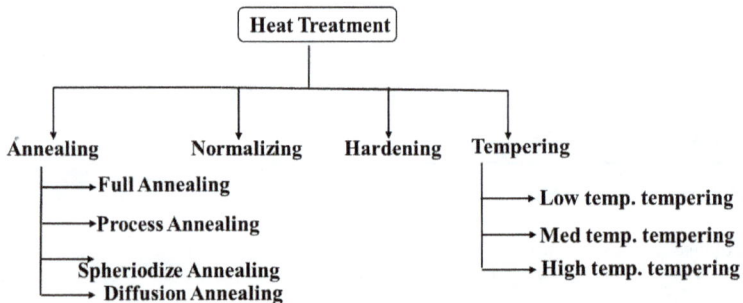

Figure 1.14: Classification of heat treatment process.

1. Annealing: Annealing is a heat treatment technique used to soften, increase ductility, and lower internal tensions in glass or metals. The material is heated to a predetermined temperature, held there for a predetermined amount of time, and then carefully cooled. This aids in achieving qualities like decreased hardness, enhanced formability, and greater machinability.

The annealing process involves three main stages:

i. **Heating**: The substance is gradually heated to a predetermined temperature, usually above its recrystallization point. The heating rate and final temperature depend on the material type and desired properties.

ii. **Holding**: The material is "soaked" or maintained at the desired temperature for a predetermined amount of time once it has been reached. This allows the internal structure of the material to homogenize, promoting uniformity in its properties.
iii. **Cooling**: The substance is then gradually cooled to room temperature. The kind of material and the required qualities determine the cooling rate. While certain substances are air-cooled or quenched in a controlled setting, others are cooled inside the furnace.

– **Classification of annealing process**
a) Full annealing: The purpose of full annealing is to soften the metal, reduce internal tension, and improve the grain structure. The heating temperature varies based on the type of steel:

– The temperature of hypoeutectoid steel is raised 30–50 °C above its highest critical temperature.
– The temperature of hypereutectoid steel is raised by 30–50 °C over its lower critical temperature (LCT).

The part is then slowly cooled in the furnace to achieve full annealing. This procedure uses full recrystallization to eliminate any structural flaws. At the high annealing temperature, the metal undergoes full crystallization, eliminating any structural defects.

In the full annealing process, the steel is heated to a predetermined temperature and maintained there for an adequate amount of time to allow internal changes in its crystal structure. Subsequently, the metal is slowly cooled in the furnace, ensuring a refined grain structure and stress relief.

b) Process annealing: Process annealing softens metals and reduces the negative effects of cold working, restoring their ductility for additional processing. During this process, the material is heated to just below the LCT (typically 550–650 °C), held at this temperature for 2–4 h depending on the metal's thickness, and then cooled slowly in the furnace or air. This treatment enhances ductility and plasticity, reduces hardness, alleviates internal stresses, and improves machinability. It is commonly utilized in the wire and sheet metal industries. Various steps of process annealing are given in Figure 1.15.

c) Spheroidize annealing: Heat treatment processes that produce spheroids are known as spheroidizing annealing. When a heat-treated product has globules of cementite (Fe_3C) within a ferrite (Fe_2O_3) matrix as shown in Figure 1.16, it is referred to as a spheroid. During spheroidizing annealing, the metal is heated just above the LCT of 730–770 °C and then cooled slowly.

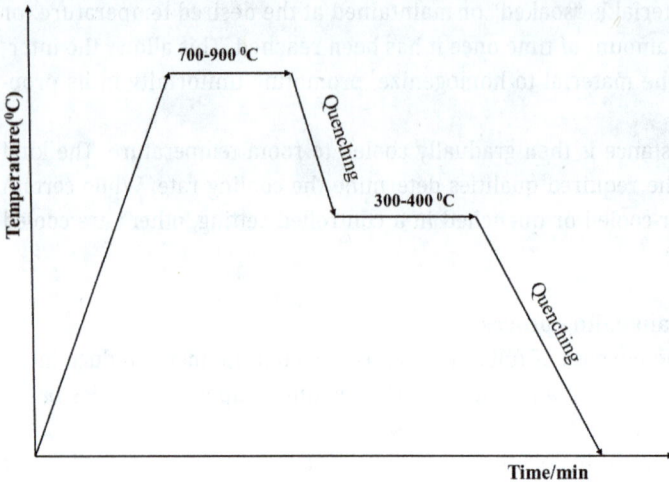

Figure 1.15: Steps in process annealing.

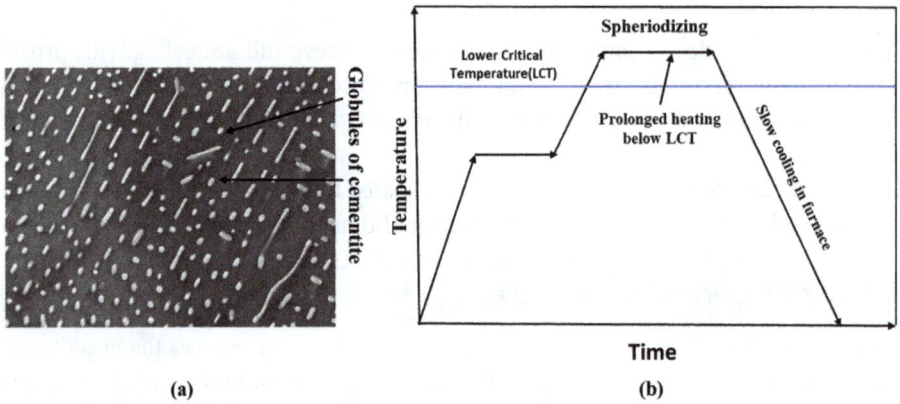

Figure 1.16: (a) Globules of cementite; (b) spheroidizing process.

d) **Diffusion annealing:** In diffusion annealing, the metal is heated to temperatures between 1,100 and 1,200 °C. At these temperatures, diffusion occurs, homogenizing the austenite grains, hence the process is also known as homogenizing annealing. Due to high temperatures, the holding time is kept short to prevent melting, followed by furnace cooling for 6–8 h down to 800 °C and then further cooling in air. This method eliminates chemical composition heterogeneity in heavy castings.

Objective of diffusion annealing
- Homogenize composition: The primary aim is to eliminate heterogeneity in the chemical composition of the metal, especially in heavy castings.
- Refine grain structure: Promotes a more uniform and refined grain structure throughout the metal.
- Reduce segregation: Minimizes segregation of alloying elements within the micro-structure.

Steps involved
- **Heating**: A high temperature, usually between 1,100 and 1,200 °C, is applied to the metal.
- **Soaking**: The austenite grains are homogenized by holding the metal at this high temperature long enough for diffusion to take place.
- **Cooling**: The metal is then cooled slowly in a furnace down to approximately 800 °C over a period of 6–8 h, followed by further cooling in air.

Uses
- Heavy castings: Used for large castings where chemical composition uniformity is critical.
- Alloy steels: Applied to alloy steels to ensure a homogeneous distribution of alloying elements.
- Complex components: Suitable for components with complex geometries that require uniform properties.

2. Normalizing: Normalizing is a heat treatment procedure that strengthens steel's mechanical qualities and grain structure. The steel is heated over its critical temperature and then allowed to cool naturally in the atmosphere.

Objective of normalizing
- Refine grain structure: Normalizing refines the grain size, resulting in improved toughness and strength.
- Enhance mechanical properties: It provides a uniform and fine-grained micro-structure, enhancing the steel's mechanical properties.
- Relieve internal stresses: It helps in relieving internal stresses that may have developed during prior processes like forging or rolling.
- Improve machinability: It makes the steel easier to machine by creating a more uniform and stable structure.

Basic steps involved
i. **Heating**: Usually, the steel is heated to a temperature of 750–980 °C, which is above the upper critical temperature.

ii. **Soaking**: This temperature is maintained for a predetermined amount of time to ensure uniform heating.
iii. **Cooling**: After then, the steel is taken out of the furnace and given time to cool in the open at room temperature.

Uses
– Structural components: Used in the production of structural steel to ensure uniform mechanical properties.
– Automotive parts: Applied to gears, shafts, and other components to enhance strength and wear resistance.
– Forgings and castings: Used for stress relief and grain structure refinement in forgings and castings.

3. Quenching, hardening, and tempering
i. Quenching: Quenching is the process of rapidly cooling a heated metal by suddenly immersing it into oil or water. This rapid cooling causes the metal to contract quickly due to the sudden heat loss. This occasionally results in the development of crack in metals.

Objective of quenching
– Increase hardness: Rapid cooling transforms the metal into a harder microstructure, often martensite in steels.
– Enhance strength: Increases the metal's resistance to wear and tensile strength.
– Refine microstructure: Creates a fine-grained microstructure, improving overall material properties.

Basic steps involved
a) **Heating**: In order to make the metal austenitic, it is heated above its critical point.
b) **Soaking**: The metal is held at this temperature to ensure uniform heat distribution.
c) **Rapid cooling**: The metal is quickly immersed in a cooling medium such as water, oil, or air to rapidly reduce its temperature.

Uses
– Tool steels: Used for making cutting tools, drills, and other high-strength applications.
– Automotive parts: Applied to gears, shafts, and other components requiring high strength and durability.
– Machinery components: Used in components that require a high level of hardness and wear resistance.

Cooling medium
- Water: Provides rapid cooling but can cause distortion and cracking due to high thermal stresses.
- Oil: Offers slower cooling than water, reducing the risk of cracking but still providing significant hardening.
- Air: Used for less severe cooling, suitable for steels that harden through slower cooling rates.

ii. Hardening: The hardening process involves heating the steel to a temperature above its critical point, maintaining it at that temperature for a sufficient period based on its thickness, and then quenching it in an appropriate medium such as water, oil, or a molten salt bath.

Objective of hardening
- Increase hardness: Enhances the hardness of the metal, making it more resistant to deformation and wear.
- Improve strength: Boosts the tensile and yield strength of the metal.
- Enhance wear resistance: Increases the durability of the metal in abrasive environments.

Basic steps involved
i. **Heating**: The temperature of the metal is raised above its critical point, transforming its structure into austenite.
ii. **Soaking**: The metal is held at this temperature to ensure uniform heat distribution.
iii. **Quenching**: The heated metal is rapidly cooled, usually by immersing it in water, oil, or air.

Uses
- Cutting tools: Used to harden blades, drills, and other cutting instruments.
- Automotive parts: Applied to gears, shafts, and other components that require high strength and wear resistance.
- Machinery components: Used for parts that need to withstand high stress and wear.

iii. Tempering: Steel that has been hardened through heating and quenching is reheated to below its LCT to transform the hard and brittle martensite into a mixture of ferrite and cementite. This process enhances toughness and ductility, although it slightly reduces hardness and strength. Table 1.7 outlines the different types of tempering—low, medium, and high—along with their respective temperature ranges and specific metallurgical purposes, such as stress relief, improved ductility, and enhanced toughness.

Table 1.7: Types of Tempering with Corresponding Temperature Ranges and Purposes.

Types	Temperature range	Purpose
Low-temperature tempering	150–250 °C	To relieve internal stresses
Medium-temperature tempering	350–450 °C	To achieve ductility and toughness
High-temperature tempering	500–650 °C	Complete elimination of internal stress

Objective of tempering
- Reduce brittleness: Lowers the brittleness introduced during the hardening process.
- Improve toughness: Enhances the toughness and impact resistance of the steel.
- Relieve stresses: Lowers any internal stresses that might have developed during the quenching process.
- Adjust hardness: Tailors the hardness of the steel to the desired level for specific applications.

Basic steps involved
i. **Heating**: The hardened reheating steel to a temperature typically between 150 and 650 °C, depending on the desired properties.
ii. **Soaking**: To allow the internal structure to stabilize, the steel is kept at this temperature for a predetermined amount of time.
iii. **Cooling**: After that, the steel is gradually cooled, usually in air to complete the process.

Uses
- Cutting tools: Tempering is used to improve the toughness of cutting tools without significantly reducing their hardness.
- Automotive components: Applied to parts like springs and gears to balance strength and durability.
- Machinery parts: Used for components that require a combination of strength and toughness such as shafts and axles.

1.3.4 Form of steel on the basis of % C and temperature

Steel can exist in different microstructural forms depending on the carbon content and the temperature. The various forms of steels on the basis of carbon % and temperature are given in Figure 1.17. The given table 1.8 summarizes the various phases found in iron-carbon alloys, detailing their carbon content, temperature stability ranges, characteristic properties, and associated microstructures.

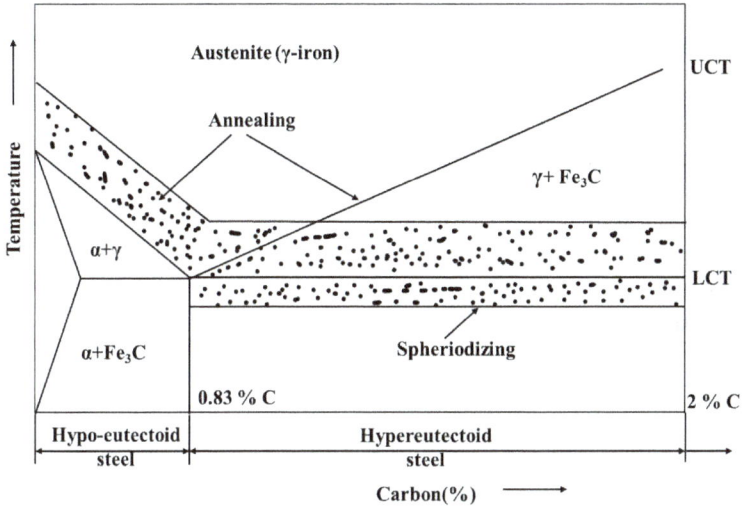

Figure 1.17: Form of steel on the basis of % C and temperature.

Table 1.8: Phases of iron-carbon alloy with carbon content, temperature, properties, and structure.

Phase	Carbon content	Temperature	Properties	Structure
Ferrite (α-Fe)	Up to 0.025% C at room temperature	Stable below 912 °C	Soft, ductile, magnetic, low strength	BCC
Austenite (γ-Fe)	Up to 2.1% C	Stable between 727 and 1,493 °C	Non-magnetic, more ductile than ferrite	FCC
Cementite (Fe₃C)	6.67% C	Forms at temperatures up to 1,493 °C	Very hard and brittle	Orthorhombic
Pearlite	Eutectoid composition at 0.8% C	Forms around 727 °C during slow cooling	Combination of ferrite and cementite, balanced strength and ductility	Lamellar (ferrite + cementite)
Bainite	Varies	Forms at temperatures between 250 and 550 °C	Higher toughness than pearlite and lower hardness than martensite	Fine plates of ferrite and cementite
Martensite	Up to 2.1% C	Forms by rapid cooling (quenching) from austenite	Very hard and brittle	BCT

1.3.5 Temperature-carbon phase diagram

The Fe-C phase diagram is a critical tool for understanding the transformations. Pressure-temperature phase diagram is shown in Figure 1.18:

i.   **Below 727 °C (eutectoid temperature):**
   –   Low carbon (up to 0.025% C): Ferrite
   –   Eutectoid composition (0.8% C): Pearlite
ii. **Between 727 and 912 °C:**
   –   Low-to-medium carbon: Mixture of ferrite and austenite.
iii. **Between 912 and 1,394 °C:**
   –   Medium to high carbon: austenite
   –   Above 2.1% C: austenite and cementite
iv. **Above 1,493 °C:**
   –   Liquid phase begins to form, leading to full melting above 1,538 °C (melting point of pure iron).

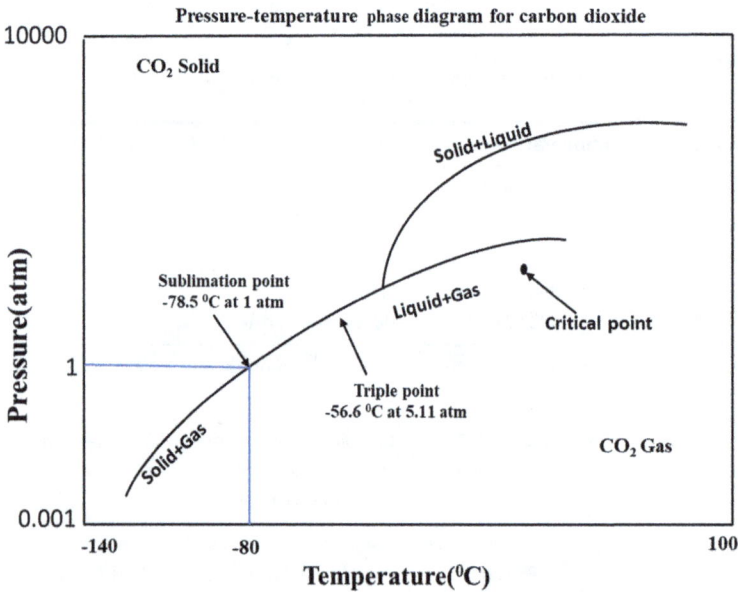

Figure 1.18: Pressure-temperature phase diagram for carbon dioxide.

1.3.6 Case hardening

A heat treatment procedure called case hardening preserves a metal's softer, ductile core while hardening its outer layer. This technique is commonly used to enhance the wear resistance and fatigue strength of components without compromising their over-all toughness.

Objective of case hardening
- Increase surface hardness: Provides a hard, wear-resistant outer layer.
- Enhance wear resistance: Improves the component's resistance to surface wear and abrasion.
- Improve fatigue strength: Increases the component's resistance to fatigue and re-peated stress.

Basic steps involved
i.   **Heating**: A high temperature is applied to the metal component.
ii. **Carbon or nitrogen enrichment**: During a carburizing or nitriding process, for exam-ple, the metal's surface is exposed to an atmosphere that is rich in carbon or nitrogen.
iii. **Diffusion**: Atoms of carbon or nitrogen permeate the metal's outermost layer.
iv. **Quenching**: To solidify the enriched surface layer, the metal is quickly cooled.

Case hardening is carried out by the following methods:

i. Carburizing
Carburizing is a process that adds carbon to a component's surface layer. This is ac-complished by heating the component to a high temperature in an environment rich in carbon, allowing carbon atoms to diffuse into the steel surface. There are three methods to accomplish carburizing:
a) **Pack carburizing**: In pack carburizing, parts that require hardening are heated in a furnace for 12–72 h at 900 °C while enclosed in a high-carbon media such as cast-iron shavings or graphite powder. Various carburizing compounds are used, with a common composition being
    - Hard charcoal: 53–55%
    - Sodium carbonate: 2–3%
    - Calcium carbonate: 3–4%
    - Barium carbonate: 10–12%

Carbon monoxide gas, a potent reducing agent, is formed at 900 °C. The carbon re-leased by the reduction reaction on the steel part's surface subsequently diffuses back into the part's surface, which is kept at 900 °C. The components are taken out and can go through another standard hardening treatment if enough carbon has been ab-sorbed (according to diffusion theory). The hardness can reach 60–65 Re, and the car-bon content on the surface can range from 0.7 to 1.2%, depending on the process cir-cumstances. The casing depth could be anywhere between 0.1 and 1.5 mm. However, one drawback of pack carburizing is the difficulty in controlling temperature uniformity and the potential for insufficient heating. The pack carburizing process and schematic diagram are shown in Figure 1.19.

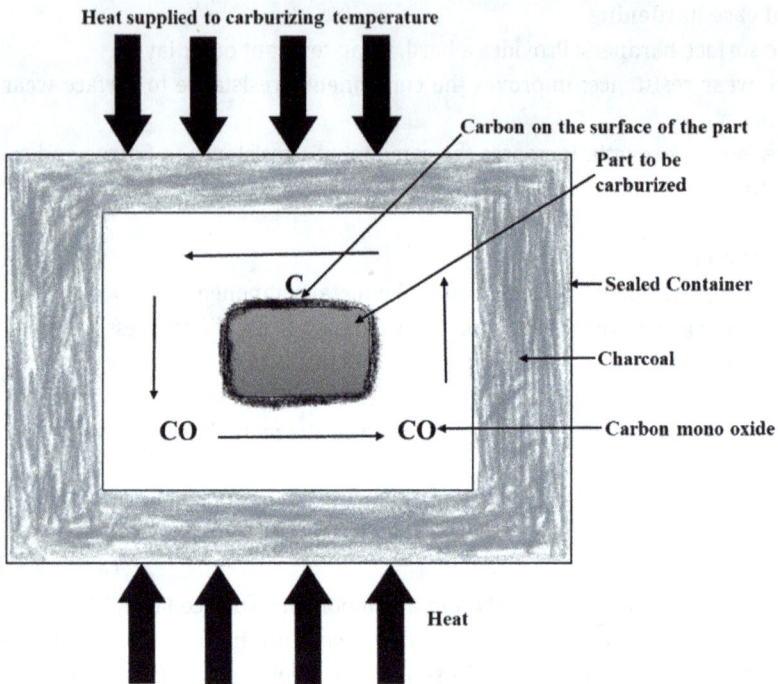

Heat supplied to carburizing temperature

Carbon on the surface of the part

Part to be carburized

C

Sealed Container

Charcoal

CO ————→ CO◄ —— Carbon mono oxide

Heat

Figure 1.19: Pack carburizing process.

b) **Gas carburizing:** Gas carburizing is a surface hardening process in which steel parts are exposed to a carbon-rich gas atmosphere at elevated temperatures, typically between 900 and 950 °C. The primary purpose of this process is to increase the carbon content on the surface layer of the steel parts, thereby enhancing their hardness and wear resistance. Gas carburizing setup is shown in Figure 1.20.

Basic steps involved

i. **Preparation**: The steel parts to be carburized are cleaned to remove any surface contaminants that might interfere with the carburizing process.

ii. **Heating**: The parts are placed in a furnace and heated to the carburizing temperature, usually between 900 and 950 °C.

iii. **Carburizing atmosphere**: A carbon-rich gas, such as methane (CH_4) or propane (C_3H_8), is introduced into the furnace. These gases decompose at high temperatures, producing carbon monoxide (CO) and hydrogen (H_2).

iv. **Diffusion**: The carbon monoxide gas acts as a reducing agent, causing carbon atoms to diffuse into the surface of the steel parts. The diffusion process forms a carbon-enriched layer on the surface.

v. **Controlled atmosphere**: The gas flow and furnace atmosphere are carefully controlled to maintain the desired carbon potential and ensure uniform carburizing.

vi. **Cooling**: After carburizing for the required time, the parts are cooled, either in the furnace or through a quenching process, depending on the desired hardness and mechanical properties.

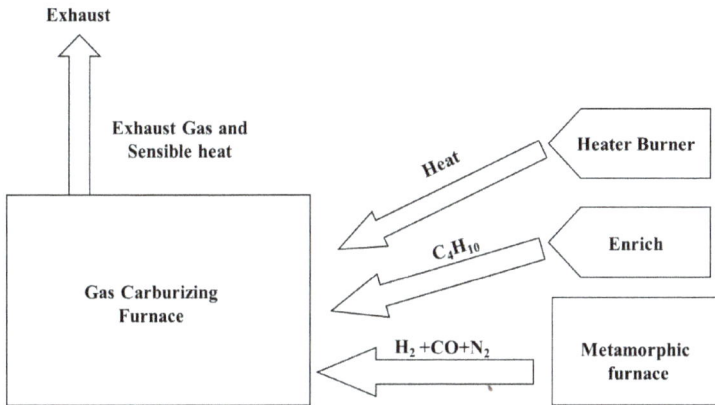

Figure 1.20: Gas carburizing setup.

c) **Liquid carburizing**: The steel parts to be case-hardened are immersed in a molten carbon-rich bath with cyanide as the primary component. However, due to safety concerns and the increased processing costs associated with using a non-toxic bath, this method is not commonly used.

ii. Nitriding: In the nitriding process, as nitrogen gas permeates steel's surface, it forms nitrides with components like vanadium, molybdenum, chromium, and aluminum. Before nitriding, the pieces are tempered and heat-treated if needed. After cleaning, the components are heated in a furnace at 500–625 °C for 10–40 h in an environment of dissociated ammonia:

$$(2NH_3 \rightarrow 2N + 3H_2).$$

Up to 0.65 mm deep, a nitride layer is formed by the diffusion of nitrogen into the steel. This nitride layer is very hard and exhibits low distortion. Further heat treatment is not applied, as it may cause cracks in the hard case.

iii. Cyaniding: Low-carbon steels are mainly case-hardened via cyaniding, a quick and effective hardening method. To get rid of any remaining cyanide, the part is heated to 871–954 °C in a sodium cyanide bath, then quenched, and cleaned in water or oil. The reaction proceeds as follows:

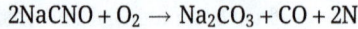

$$2NaCN + 2NaCNO \rightarrow 2NaCNO$$

$$2NaCNO + O_2 \rightarrow Na_2CO_3 + CO + 2N$$

$$2CO \rightarrow CO_2 + C$$

The released carbon is absorbed by the part's surface, creating a hard shell with a thickness between 0.25 and 0.75 mm within 20–30 min. This thin hardened layer is harder than that produced by traditional case hardening treatments, with minimal risk of distortion. Typical applications include bolts, nuts, screws, and small gears.

1.4 Non-ferrous metals

1.4.1 Aluminum

- **Process details:** extraction of aluminum from bauxite
  - Raw material: Bauxite contains impurities like iron, silicon, and titanium oxides.
  - Purification: Bauxite undergoes a purification process to remove impurities.
  - Fusion: The bauxite is fused in an electric furnace.
  - Addition of carbon: Carbon is added to eliminate impurities by forming a removable sludge.
  - Formation of aluminum oxide: After impurity removal, aluminum oxide remains.
  - Electrolytic reduction: Aluminum is separated from aluminum oxide using an electrolytic bath.
  - Collection of pure aluminum: Pure aluminum collects at the bottom of the bath.
  - Extraction: Aluminum is extracted at regular intervals.
- **Properties:** High electrical and heat conductivity, resistance to corrosion, high ductility, and light weight.
- **Uses:** It is used for conductors and bus bars, domestic utensils, heat-conductive applications, and containers for chemical industries. It can be easily shaped, rolled, drawn, and forged.

1.4.2 Copper

- **Process details:** extraction of copper
  - Raw material: Copper is primarily available as copper pyrites.
  - Roasting: The ore undergoes roasting to remove water, CO_2, and sulfur.

- Melting: The roasted ore is melted in a reverberatory furnace.
- Addition of silica: Silica is added to eliminate impurities like iron and alumina.
- Tapping of molten metal: The molten metal is tapped and sent to a converter.
- Air blowing: Air is forced through the molten metal to remove remaining impurities, producing crude blister copper.
- Electrolytic refining: The crude copper undergoes electrolytic deposition, refining it to 99.9% purity.
- Final casting: The purified copper is cast into suitable shapes for use.
- **Properties:** High electrical and heat conductivity, good corrosion resistance and high ductility.
- **Uses:** Used as electrical conductors in the form of wire, sheet, and contacts, it is also utilized in heat exchangers and heating vessels. Additionally, it serves as a base coating on steel before nickel and chromium plating. It can be cold worked, rolled, and drawn.

1.4.3 Magnesium

- **Process details:** extraction of magnesium
  - Raw material: Magnesium is primarily obtained from minerals like dolomite ($MgCO_3 \cdot CaCO_3$), magnesite ($MgCO_3$), and seawater ($MgCl_2$).
  - Calcination: The ore is heated to decompose magnesium carbonate into magnesium oxide (MgO) and carbon dioxide (CO_2).
  - Reduction process:
    - Magnesium oxide (MgO) is mixed with a reducing agent like ferrosilicon (FeSi).
    - The mixture is heated in an electric furnace at high temperatures (1,200–1,400 °C) under a vacuum to reduce MgO into magnesium vapor.
  - Condensation: The magnesium vapor is cooled and condensed into solid magnesium.
  - Electrolytic extraction (alternate method):
    - If using seawater, magnesium chloride ($MgCl_2$) is extracted and subjected to electrolysis.
    - Electrolysis separates magnesium metal at the cathode and chlorine gas at the anode.
  - **Purification and casting:** The crude magnesium is refined and cast into ingots or other suitable shapes for industrial use.
- **Properties:** The lightest metal can be die-cast, producing pressure-tight castings with a good finish. It is easily formed, drawn, forged, and precisely machined. Alloying with aluminum, zinc, and manganese enhances its casting and extrusion properties.
- **Uses:** Motor car gear box, differential housing with and portable tool.

1.4.4 Zinc

- **Process details: extraction of zinc**
  - Ore concentration: The zinc ore is concentrated to remove impurities.
  - Retort feeding: The concentrated ore is fed into a retort along with a carbonaceous material.
  - Heating process: Multiple retorts are placed in a furnace, and the temperature is raised to 1,100 °C.
  - Zinc vaporization: At high temperatures, zinc vaporizes.
  - Condensation: The zinc vapor is rapidly cooled to condense into zinc powder.
- **Properties:** High corrosion resistance, low melting point, high fluidity.
- **Uses:** It is utilized as a protective coating on iron and steel. Its low melting point and high fluidity make it ideal for die casting.

1.4.5 Lead

- **Process details: extraction of lead**
  - Raw material: Lead ores mainly occur as oxides and high-quality sulfides, often containing impurities like iron, copper, and zinc.
  - Ore concentration: The ore is concentrated to remove unwanted materials.
  - Addition of flux: Silica and lime are added as flux.
  - Smelting in blast furnace: The ore and flux mixture is heated in a small blast furnace at 1,010 °C.
  - Separation of impurities: Impurities form a liquid slag, while lead melts and collects at the bottom.
  - Removal of molten lead and slag: The molten lead and slag are periodically removed from the furnace.
  - Further refining: The crude lead undergoes additional purification in a reverberatory furnace under an oxidizing environment to remove any remaining contaminants.
- **Properties:** Soft, heavy, and pliable, with good resistance to corrosion and chemicals alloyed with free-cutting qualities and brass.
- **Uses:** It is employed for water pipes and roof protection as well as in constructing acid bath containers for the alloyed chemical industry. This metal is easily shaped and worked and is utilized in the production of soft solder and plumbing solder.

1.4.6 Tin

- **Process details: extraction of tin**
    - Raw material: Tin is naturally available as cassiterite (SnO_2), often containing impurities like copper, iron, lead, antimony, bismuth, and zinc.
    - Ore concentration: The ore is concentrated to remove unwanted materials.
    - Roasting: The concentrated ore is roasted to eliminate excess arsenic and sulfur.
    - Smelting in reverberatory furnace: The roasted ore is heated in a reverberatory furnace with anthracite (a carbon source).
    - Reduction reaction: The anthracite reduces tin oxide (SnO_2) to metallic tin.
    - Separation of tin: The molten tin sinks to the bottom of the furnace.
    - Tapping: The molten tin is tapped at regular intervals.
    - Refining: The crude tin can undergo further refining to improve purity.
- **Properties:** Soft, good plasticity, easy to work with, rolled in thin foils, and unable to be drawn because of its low strength. It also has good resistance to acid corrosion.
- **Uses:** Used as an alloying in soft element solders, bronze, and bearing metals.

1.4.7 Nickel

- **Process details: extraction of nickel**
    - Roasting: The ore is roasted to remove volatile impurities.
    - Smelting in blast furnace: The roasted ore is melted in a small blast furnace.
    - Addition of flux: Limestone and quartz are added as flux to form slag and remove impurities.
    - Tapping of molten nickel: The crude molten nickel is periodically tapped from the furnace.
    - Refining in Bessemer converter: The crude nickel is further refined in a Bessemer converter.
    - Treatment with sulfuric acid: The refined nickel is treated with sulfuric acid to dissolve and remove copper as copper sulfate.
    - Separation of pure nickel: The pure nickel is extracted after the removal of copper.
- **Properties:** Good resistance to both acid and basic corrosion, excellent tensile strength, and ease of hot and cold working.
- **Uses:** Find wide use in and basic food processing, plated on steel to provide corrosion resistant coating. Important alloying element for steel.

1.5 Non-ferrous alloys

Non-ferrous metals, due to their high cost and low strength, are rarely used in their pure form. However, their alloys possess excellent physical and mechanical properties, making them suitable for industrial applications. These alloys offer advantages such as a high strength-to-weight ratio, ease of fabrication, good machinability, high corrosion resistance, good castability, and an attractive appearance. Given table 1.9 provides an overview of widely used non-ferrous alloys, highlighting their key mechanical and physical properties that make them suitable for various industrial and engineering applications.

1.5.1 Brass alloy

Brass is an alloy of copper (55–90%) and zinc (10–45%), with small additions of lead, tin, or aluminum for enhanced properties. It has excellent corrosion resistance, machinability, and thermal conductivity. Brass is widely used in electrical fittings, musical instruments, plumbing, and decorative items due to its strength and aesthetic appeal.

Table 1.9: Properties of commonly used non-ferrous alloys.

Alloy	Composition	Properties	Applications
Brass	Alpha beta brass: comprises between 33% and 46% zinc. Ideal for heated work Typical brasses utilized in engineering	Alpha brasses are ductile, cold worked, rolled, or drawn into wire and deep drawn to tube	Good for hot working Common brasses used in engineering application
	Cartridge brass: 70% Cu + 30% Zn	At high temperatures, it loses strength but becomes extremely flexible	Cartridge cases, head lamp reflector, drawn tube
	Muntz metal: 60% Cu + 40% Zn	Ductile	Casting of pulp parts and valve taps
	Naval brass: 60% Cu + 39% Zn + 1% tin	General-purpose alloy	Used for ship fittings that are cast and forged
	Silicon brass: 80% Cu + 16% Zn + 4% Si	Resistance to seawater corrosion	Used for refrigerators and fire extinguisher gravity die cast shell
	Beta brass: 50% Cu + 50% Zn	Can be easily sand, gravity die cast	Brazing solder
	Delta brass: 60% Cu + 37% Zn + 3% iron	Hard and brittle and can be soften	Used at place of steel casting

Table 1.9 (continued)

Alloy	Composition	Properties	Applications
	Free cutting brass: 60% Cu + 37% Zn + 3% Pb	Can be forged, rolled, extruded, and cast	Used for machining work
	Gilding brass: 15% Zn + 85% Cu	Cast, forged, stamped	Jewelry, decorative, and ornament item
	Admiralty brass: 70% Cu + 29% Zn + 1% tin	Possess strong cold-working skills. Cold-workable and resistant to corrosion from seawater	Ship fitting like bolts, nuts, and washer

1.5.2 Bronze

They are alloys of Cu and tin (up to 8%). Bronzes are easier to cold work, roll, form, and draw than brass, and they are more resistant to corrosion and mechanical stresses. The typical compositions of various bronze alloys, as shown in Table 1.10, along with their key properties and common applications, illustrating their usefulness in engineering, electrical, and decorative fields.

Table 1.10: Composition, properties, and applications of bronze alloys.

Name	Composition	Properties	Applications
Aluminum bronze	9–12% Al and up to 6% iron and Ni	High strength, corrosion, and wear resistance	Used in heavy-duty sleeve bearing machine tool ways
Silicon bronze	0.5–1% iron, 0.25–1.25% Mn, 1–4% Si, and the remainder copper	High resilience to corrosion, strength, and toughness. Hot work is possible	Used for boiler parts, tanks, marine hardware, etc.
Manganese bronze	55–60% Cu, 38–42% Zn, up to 1.5% tin, up to 2% iron, up to 1.5% Al and up to 3.5% Mn	Excellent and good rudder parts show corrosion resistance. Poor response to cold working	Ship propeller, rudders, etc. Components that need to be extremely strong and resistant to corrosion
Gun metal	2–11% tin Provides high bearing capacity and 1–10% zinc	Provides high pressure bearing	Valves, pipe fitting, and pumps

1.5.3 Aluminum alloys

Aluminum alloys contain aluminum mixed with elements like copper, magnesium, silicon, and zinc for strength and durability. They are lightweight, corrosion-resistant, and highly machinable. Used in aerospace, automotive, construction, and packaging

Table 1.11: Comparison of various aluminum alloys.

Name	Composition	Properties	Applications
Duralumin	0.5% Mg, 0.5% Mn, 4% Cu, and the remaining aluminum	Ductile, easily soft, rolled, forged, extruded, or drawn	Because of their high strength and low weight, aircraft
Aluminum casting alloy	1% iron, 8% copper, 90% aluminum, and 1% silicon	Good strength, hardness, and machinability	Used to gravity die cast and sand cast
Y-alloy	93% Al, 4% Cu, 2% Ni, and 1% Mg	Maintain strength at higher temperature	Piston engine of IC

industries, they offer excellent strength-to-weight ratio and thermal conductivity. Table 1.11, compares different types of aluminum alloys, highlighting their compositions, mechanical properties, and suitability for various industrial applications.

1.5.4 Nickel alloys

Nickel alloys consist of nickel mixed with elements like chromium, iron, copper, and molybdenum, enhancing strength, corrosion, and heat resistance. They are used in aerospace, chemical processing, marine, and power generation industries for applications requiring durability, high-temperature stability, and oxidation resistance. A comparison of various nickel alloys, as shown in Table 1.12, highlighting their compositions, properties, and typical industrial applications.

Table 1.12: Comparison of various nickel-based alloys.

Name	Composition	Properties	Applications
German silver	60% Cu, 20% Ni, and 20% Zn alloy	Difficult to work for jewelry purpose, harder exposure to air, it turns green	Used in flatware industry (electroplated nickel silver)
Constantan	45% Ni, 55% Cu. Unaffected by temperature variation	High-specific resistance	Register, thermocouple, Wheatstone bridge, etc.
Monel metal	Among other metals, there are 68% Ni, 30% Cu, 1% iron, and trace amounts of Mn	Superior strength and resistance to corrosion in alkalis, seawater, hydrofluoric acid, sulfuric acid, and other chemicals	Utilized in chemical processing, maritime engineering, valves, pumps, and other equipment
Inconel	80% Ni, 14% Cr and 6% known for their iron, resistance to oxidation	Maintain structural integrity in high-temperature environment	Chemical industry

Table 1.12 (continued)

Name	Composition	Properties	Applications
Incoloy	Iron makes up the remaining 42% Ni, 13% Cr, 6% Mo, high 2.4% Ti, and 0.04% C	High-temperature resistance	Used as a high-temperature alloy
K-Monel	Similar to Monel in addition to 3–4% Al	Better mechanical properties than Monel	Similar applications as Monel metal
Nimonic	80% Ni, 20% Cr	High strength can operate under intermittent heating and cooling condition	Gas turbine engine

1.5.5 Bearing metal

Bearing metal, also known as Babbitt metal, is an alloy primarily composed of tin, lead, copper, and antimony. It offers excellent wear resistance, low friction, and high load-bearing capacity. Used in engine bearings, turbines, and heavy machinery, it ensures smooth operation and reduces wear between moving parts.

Bearing metal has the following properties:
1. Enough compressive strength
2. Good plasticity
3. Reduced coefficient of friction to prevent overheating

Some important bearing metal is followed:

Lead alloy (40% Pb and 60% Cu): It may be cast as a thin shell **Babbit metal** (85% Tin, 10% Sb, and 5% Cu) used for heavy-duty vehicle.

Questions

Long answer-type questions
1. Describe the fundamental differences between ductile and brittle materials. Discuss the implications of these differences in the context of engineering applications.
2. Explain the process of strain hardening in metals. How does it affect the mechanical properties of a material? Provide examples of materials that undergo significant strain hardening.
3. Discuss the mechanisms of creep in materials. What are the different stages of creep and how do they impact the long-term performance of engineering components?
4. Compare and contrast the mechanical properties of ferrous and non-ferrous alloys. Include examples of specific materials and their applications in engineering.
5. Explain the concept of fatigue in materials. What factors influence the fatigue life of a material and how can engineers design components to minimize fatigue failure?

6. Discuss the significance of the stress-strain curve in understanding the mechanical behavior of materials. Explain the key points on the curve and their relevance to material selection and design.
7. How do temperature and environmental conditions affect the mechanical properties of materials? Provide examples of materials that are particularly sensitive to these factors and discuss how they are used in engineering.
8. Describe the different types of composite materials and their mechanical properties. How do the properties of composite materials differ from those of their individual constituents?
9. Explain the concept of hardness in materials. How is hardness measured and what is its significance in engineering applications? Discuss the various hardness testing methods.
10. Discuss the impact of grain size on the mechanical properties of metals. Explain the Hall-Petch relationship and how grain size can be controlled in engineering materials.
11. Discuss the various types of carbon steel, their compositions, and their respective mechanical properties. How do these properties influence their applications in engineering?
12. Compare and contrast the microstructure and properties of gray cast iron, white cast iron, ductile cast iron, and malleable cast iron. How do these differences influence their usage in industrial applications?
13. Describe the methods used to improve the mechanical properties of cast iron. What are the advantages and disadvantages of these methods? Provide examples of applications where these improved cast irons are utilized.
14. What are alloy steels, and how do alloying elements such as chromium, nickel, molybdenum, and vanadium affect the properties of steel? Provide examples of specific alloy steels and their typical applications.
15. Discuss the role of non-ferrous materials in engineering. Compare their properties with ferrous materials and explain why they are chosen for specific applications. Include examples of common non-ferrous materials such as aluminum, copper, titanium, and their alloys.
16. Analyze the environmental and economic implications of using carbon steel, cast iron, alloy steel, and non-ferrous materials in engineering. How do factors like recyclability, material scarcity, and production costs impact material selection?
17. Explain the process of annealing in detail. How does it affect the microstructure and mechanical properties of metals? Provide examples of materials commonly subjected to annealing and the specific changes they undergo.
18. Compare and contrast annealing and normalizing. Discuss the similarities and differences in terms of their objectives, process parameters, and the resulting changes in the properties of the treated metals.
19. Describe the tempering process. How does tempering affect the hardness and toughness of a material? Illustrate your answer with the tempering of martensitic steel and the typical tempering curves.
20. What is the significance of hardening in heat treatment? Explain the different methods of hardening (such as quenching, induction hardening, and flame hardening) and the factors that influence the choice of hardening technique.
21. Case hardening is often used to enhance surface properties of metals. Describe the various case hardening techniques (carburizing, nitriding, carbonitriding, etc.) and explain how they improve wear resistance and fatigue strength of components.
22. Discuss the microstructural changes that occur during the hardening and tempering processes. How do these changes correlate with the mechanical properties such as hardness, ductility, and tensile strength?
23. Explain the role of cooling rate in the heat treatment processes of annealing, normalizing, and hardening. How does the cooling rate affect the final microstructure and mechanical properties of the treated metal?
24. In what ways does the heat treatment of alloy steels differ from that of plain carbon steels? Discuss the challenges and considerations involved in the heat treatment of alloy steels, including the effects of alloying elements on the heat treatment process.
25. Define the term "heat treatment cycle" and describe a typical heat treatment cycle for the production of a high-performance steel component. Include steps like heating, soaking, and cooling, and explain the importance of each step.

26. Discuss the environmental and economic impacts of heat treatment processes. How can advancements in heat treatment technology contribute to sustainable manufacturing practices? Provide examples of innovations or techniques that reduce the energy consumption and environmental footprint of heat treatment operations.
27. Discuss the properties and applications of aluminum alloys in the automotive and aerospace industries. How do different alloying elements affect the properties of aluminum?
28. Explain the significance of copper and its alloys in electrical and thermal applications. What are the key properties that make copper alloys suitable for these purposes?
29. Describe the process of titanium extraction and its alloying techniques. How do titanium alloys compare to steel in terms of strength-to-weight ratio, corrosion resistance, and applications?
30. Examine the role of magnesium alloys in the manufacturing of lightweight structures. What challenges are associated with the use of magnesium, and how are these addressed in practical applications?
31. Analyze the benefits and limitations of using nickel-based superalloys in high-temperature environments. What are the common applications of these superalloys, and how do they perform under extreme conditions?
32. Discuss the various methods of heat treatment applied to non-ferrous alloys. How do these treatments enhance the mechanical properties and performance of the materials? Provide specific examples.
33. Investigate the corrosion resistance properties of stainless steel compared to other non-ferrous alloys. What are the mechanisms of corrosion in these materials, and how can they be prevented or mitigated?
34. Explore the use of non-ferrous alloys in medical applications, particularly in implants and surgical instruments. What properties are essential for these materials, and how are they tested for biocompatibility and durability?
35. Discuss the impact of alloying elements on the phase diagrams of non-ferrous materials. How do phase diagrams help in understanding the solidification and thermal treatment processes of these alloys? Provide examples of common phase diagrams for non-ferrous alloys.

Short answer-type questions

1. Define hardness and explain how it is measured.
2. What is tensile strength and how is it different from compressive strength?
3. Describe ductility and provide an example of a ductile material.
4. What is malleability, and how does it differ from ductility?
5. Explain the concept of elasticity. How is it different from plasticity?
6. Define plasticity and give an example of a material that exhibits high plasticity.
7. What is fatigue in materials, and what are the common signs of fatigue failure?
8. Describe the phenomenon of creep and the conditions under which it typically occurs.
9. How is the hardness of a material tested using the Rockwell hardness test?
10. What factors influence the fatigue life of a material in engineering applications?
11. What is the primary difference between carbon steel and alloy steel?
12. List three common applications of gray cast iron and explain why it is suitable for these applications.
13. What is the main characteristic that differentiates stainless steel from other types of steel?
14. Describe the purpose of adding chromium to alloy steels.
15. How does the carbon content affect the hardness and ductility of carbon steel?
16. What is the main advantage of using ductile cast iron over gray cast iron in engineering applications?
17. Name two non-ferrous materials commonly used in aerospace applications and briefly explain why they are preferred.
18. What are the key benefits of using aluminum alloys in automotive manufacturing?
19. Explain why copper is often used in electrical applications.
20. Identify one primary use for titanium alloys and explain the property that makes titanium suitable for this use.

21. What is the primary purpose of annealing in heat treatment?
22. How does normalizing differ from annealing in terms of cooling process and resulting microstructure?
23. What is the main objective of tempering in the heat treatment of steel?
24. Describe the process of hardening and its effect on the mechanical properties of a material.
25. What is case hardening, and which applications typically require this process?
26. In process annealing, what specific purpose does it serve in comparison to full annealing?
27. Explain the term "full annealing" and the typical steps involved in this process.
28. What is gas carburizing, and how does it differ from liquid carburizing in terms of procedure and application?
29. Describe the role of carbon in the process of gas carburizing.
30. What are the advantages of liquid carburizing over other surface hardening techniques?
31. What are non-ferrous metals? Provide two examples.
32. Describe the primary advantages of using aluminum in engineering applications.
33. What are the main characteristics of copper that make it suitable for electrical wiring?
34. Explain the significance of magnesium alloys in aerospace engineering.
35. What is the role of zinc in galvanizing steel?
36. How does the presence of silicon in aluminum alloys affect their properties?
37. Why is titanium often used in medical implants and aerospace components?
38. What are some common applications of nickel-based superalloys?
39. How do non-ferrous metals typically respond to corrosion compared to ferrous metals?
40. What are the environmental considerations associated with recycling non-ferrous metals?

Questions asked in competitive examinations

1. Which of the following materials has the highest ductility?   **[GATE 2020]**
   (A) Cast iron   (B) Mild steel   (C) Tool steel   (D) High carbon steel
2. Gray cast iron is primarily characterized by   **[ESE 2019]**
   (A) High ductility
   (B) High tensile strength
   (C) Graphite flakes in its microstructure
   (D) High toughness
3. Which of the following has the highest carbon content?   **[GATE 2018]**
   (A) Wrought iron   (B) Mild steel   (C) High carbon steel   (D) Cast iron
4. The hardness of steel is increased by   **[ESE 2017]**
   (A) Annealing   (B) Normalizing   (C) Hardening   (D) Tempering
5. Which of the following heat treatment processes is used to soften a hardened material? **[GATE 2016]**
   (A) Hardening   (B) Tempering   (C) Annealing   (D) Normalizing
6. Spheriodize in steels is obtained by   **[ESE 2015]**
   (A) Quenching
   (B) Rapid cooling
   (C) Prolonged heating at a temperature below the eutectoid
   (D) Cold working
7. The percentage of carbon content in mild steel is typically: **[GATE 2014]**
   (A) 0.01–0.1%   (B) 0.1–0.3%   (C) 0.3–0.6%   (D) 0.6–0.9%
8. Which of the following statements is true for ductile iron?   **[ESE 2013]**
   (A) It has high ductility due to the presence of graphite flakes.
   (B) It is produced by adding magnesium to molten iron.
   (C) It is more brittle than gray cast iron.
   (D) It has low impact resistance.

9. The main alloying element in high-speed steel (HSS) is   **[GATE 2012]**
   (A) Manganese   (B) Chromium   (C) Tungsten   (D) Vanadium
10. The critical temperature above which austenite is stable in carbon steel is   **[ESE 2011]**
    (A) 723 °C   (B) 910 °C   (C) 1,150 °C   (D) 1,350 °C
11. Which of the following heat treatment processes is used to increase the toughness of hardened steel?   **[GATE 2020]**
    (A) Annealing   (B) Normalizing   (C) Tempering   (D) Hardening
12. Which of the following is NOT a surface hardening process?   **[ESE 2019]**
    (A) Nitriding   (B) Case hardening   (C) Carburizing   (D) Tempering
13. In the heat treatment process, which of the following is typically used to relieve internal stresses?   **[GATE 2018]**
    (A) Quenching   (B) Tempering   (C) Annealing   (D) Hardening
14. During the process of normalizing, the steel is: **[ESE 2017]**
    (A) Heated to a temperature above the critical range and cooled in still air
    (B) Heated to a temperature below the critical range and cooled in still air
    (C) Heated to a temperature above the critical range and cooled in furnace
    (D) Heated to a temperature below the critical range and cooled in furnace
15. Which heat treatment process is used to soften the metal and improve machinability?   **[GATE 2016]**
    (A) Hardening   (B) Normalizing   (C) Annealing   (D) Quenching
16. Which of the following heat treatment processes involves heating a material to a high temperature followed by rapid cooling in water or oil?   **[ESE 2015]**
    (A) Tempering   (B) Annealing   (C) Hardening   (D) Normalizing
17. What is the main advantage of gas carburizing over liquid carburizing?   **[GATE 2014]**
    (A) Higher hardness
    (B) Lower cost
    (C) Better control of case depth
    (D) Faster processing time
18. Which of the following processes is used to improve the wear resistance of low carbon steel?   **[ESE 2013]**
    (A) Annealing   (B) Normalizing   (C) Carburizing   (D) Tempering
19. What is the main purpose of full annealing in the heat treatment of steel?   **[GATE 2012]**
    (A) To increase hardness
    (B) To relieve internal stresses
    (C) To increase strength
    (D) To refine grain structure
20. Which of the following is a characteristic feature of process annealing?   **[ESE 2011]**
    (A) It is performed at a temperature above the upper critical temperature.
    (B) It involves slow cooling in a furnace.
    (C) It is used primarily to soften work-hardened materials.
    (D) It is used to harden materials.

Answers
1(B)   2(C)   3(D)   4(C)   5(C)   6(C)   7(B)   8(B)   9(C)   10(B)   11(C)   12(D)   13(C)   14(A)   15(C)   16(C)
17(C)   18(C)   19(B)   20(C)

Chapter 2
Metal forming processes and applications

2.1 Introduction to hot and cold working

1. **Recrystallization temperature:** When heat is supplied to a metal for a certain period, it eventually reaches a stage where the existing grains of the metal are destroyed, and new grains begin to form. This process is known as recrystallization. The specific temperature at which this transformation occurs is referred to as the recrystallization temperature. During recrystallization, the metal undergoes significant changes in its microstructure, leading to the formation of new, strain-free grains that replace the deformed ones, thereby restoring the metal's ductility and reducing internal stresses. The recrystallization temperature is a critical parameter in materials science, as it influences the properties and performance of the metal in various applications.

2. **Hot working:** When a metal is plastically deformed at a temperature above its recrystallization temperature, the process is known as hot working. During hot working, the metal's grain structure can be effectively controlled and refined. This is because the high temperature allows the metal to recrystallize simultaneously as it is being deformed, resulting in the continuous formation of new, strain-free grains. This process not only enhances the metal's ductility but also helps in achieving desired mechanical properties by manipulating the grain size. Controlling the grain size during hot working is crucial, as finer grains can improve the metal's strength, toughness, and resistance to deformation.

3. **Cold working:** When a metal is plastically deformed at a temperature below its recrystallization temperature, the process is known as cold working. There are several important differences between hot-worked and cold-worked products. For example, the dimensional accuracy of hot worked products is generally lower due to the effects of thermal expansion and contraction during the process. Additionally, during heating, a layer of oxide forms on the surface of the metal, leading to a rough surface on the worked products. In contrast, cold working, performed at lower temperatures, results in higher dimensional accuracy and a smoother surface finish. Cold-worked metals also tend to have increased hardness and strength due to work hardening though they may exhibit reduced ductility compared to hot worked metals.

The differences between hot working and cold working processes are clearly outlined in Table 2.1, which compares them based on factors such as working temperature, mechanical properties, and surface finish. Further insight into the practical aspects of these processes is provided in Table 2.2, which summarizes their respective advantages and limitations, helping to understand their suitability for different manufacturing applications.

https://doi.org/10.1515/9783112205891-002

Table 2.1: Hot working vs. cold working.

Hot working	Cold working
Process carried out above recrystallization temperature.	Process carried out below recrystallization temperature.
No strain hardening occurs in hot working.	Material become brittle due to hardening.
Less stress is required to produce the deformation due to high temperature.	Stress required to deformation is increases suddenly with the amount of deformation.
For large deformation, hot working is preferred.	For small deformation cold working is preferred.
Low energy required even for bulk deformation.	Higher energy is required even for less deformation.
Poor surface finish due to scaling and oxidation.	Good surface finish is obtained.
Light equipment needed.	Heavy equipment needed.
Difficult to handle manually due to high temperature.	Easy to handle manually.
Refine the grain structure.	No refinement occurred in cold working.
Some defects like internal porosity, blow holes, cracks etc. appeared during hot working.	Only crack formation due to excessive load.
Hot rolling, hot forging, hot extrusion, spinning, pipe welding, hot drawing, hot shearing, etc.	Cold extrusion, cold rolling, press working, etc.

Table 2.2: Advantages and limitations of hot working and cold working.

Hot working	Cold working
Hot working enhances the mechanical properties of the metal.	Achieves a superior surface finish.
It refines the grain structure, leading to improved material performance.	Ensures greater dimensional accuracy.
It improves the properties compared to those of cast products.	Enhances mechanical properties.
It alleviates residual stresses developed during welding and casting.	Results in a loss of ductility.
A layer of oxide forms on the surface, resulting in a rough finish.	Induces strain hardening in cold worked products.
Surface decarburization occurs during the process.	Requires higher force to deform the metal.
Dimensional accuracy is reduced due to thermal expansion and contraction.	Increases yield strength and hardness.

Table 2.2 (continued)

Hot working	Cold working
Increased ductility and toughness are achieved through the process.	Reduces the size of defects and imperfections.
Enables the production of larger and more complex shapes.	Leads to the formation of finer microstructures.
Helps in homogenizing the chemical composition of the metal.	May introduce residual stresses.
	Improves surface quality with less oxidation compared to hot working.

General classification of metal forming processes is given in Figure 2.1.

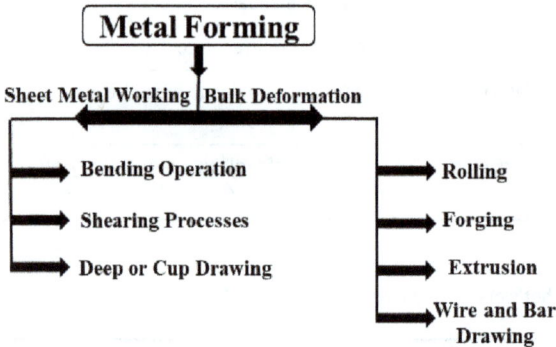

Figure 2.1: Classification of metal forming process.

– Material properties in metal forming

For effective metal forming, it is essential to have properties such as low yield strength and high ductility. As the working temperature increases, these properties are influenced: ductility improves and yield strength decreases, facilitating easier deformation. Additionally, other crucial factors to consider include strain rate and friction. A higher strain rate can increase the material's resistance to deformation, while friction between the metal and forming tools affects the efficiency and quality of the process. Proper lubrication and control of strain rate are therefore important to optimize metal forming operations and achieve desired results. Temperature control also plays a critical role in maintaining the balance between these factors to ensure the best outcomes in the formed products.

2.2 Basic metal forming process

2.2.1 Forging process

Forging is a metalworking process (see Figure 2.2), where heated metal is shaped into strong parts under high pressure, known as forgings. The heated metal is placed on a mold (die) and then pressed, pounded, or squeezed using a press ram or hammer. This pressure makes the hot, soft metal take the shape of the mold cavity. The pressure is highest when the two halves of the mold are fully closed together. Excess metal flows out at the seam of the mold, forming a ridge called flash, which is removed later in the finishing process. A proper lubricant is applied between the mold surfaces and the metal to prevent sticking, reduce wear on the mold, and act as a thermal insulator.

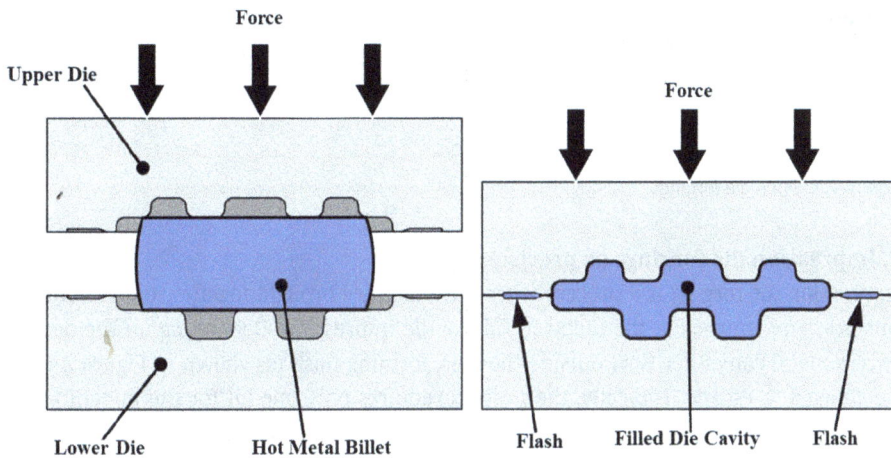

Figure 2.2: Forging operation.

– **Types of forging process**
1. Open die forgings or hand forgings
In open die forging, the metal billet is plastically deformed between an anvil and a hammer, as illustrated in Figure 2.3. In this process, the flow of the metal is not confined by any mold, allowing the metal to spread freely. The desired shape is achieved by skillfully positioning and repositioning the workpiece between repeated blows of the hammer and the anvil. This method is typically performed using a hand hammer; hence it is also known as hand forging.

During open die forging, the blacksmith or operator manually manipulates the billet to gradually shape it into the desired form. This process requires a high level of skill and experience to control the deformation and ensure uniformity in the final product. Open die forging is often used for creating large or custom-shaped metal parts and is particularly suitable for producing simple shapes or preforms for further processing.

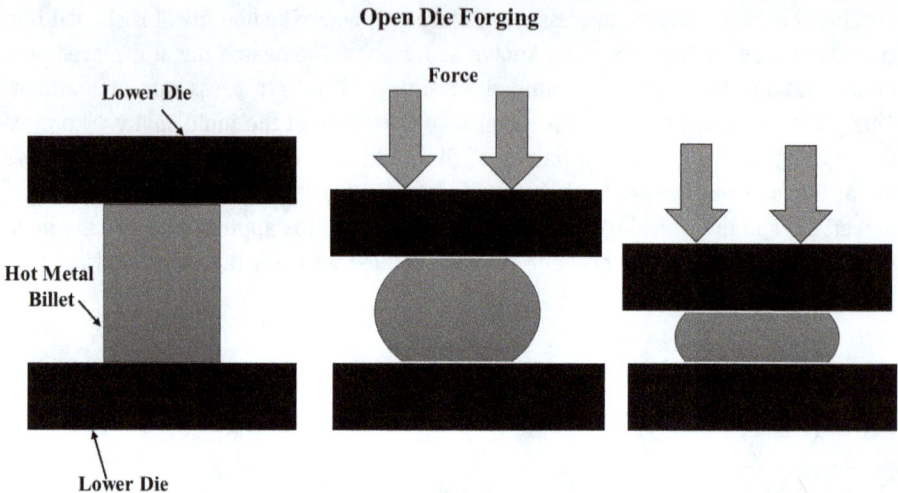

Figure 2.3: Open die forging.

2. Impression die forgings or precision forgings

Impression die forging is a process where two dies are brought together under continuous pressure, squeezing the metal to fill the die impression. Depending on the design, excess metal can either flow outside the dies, forming flash (as shown in Figure 2.4), or be allowed to escape. Typically, the process requires only one hit for the metal to take the shape of the die cavity. Impression die forging relies heavily on the pressure applied by the dies, with minimal involvement from the operator using a heavy hammer. Schematic sketch of impression die forging showing flash formation is shown in Figure 2.4.

Impression die forging, where metal is compressed between shaped dies, forcing it to take the die's form, enhancing precision and strength in the final product.

3. Closed die forging/flash less forging

In closed die forging (see Figure 2.5), the metal flow is completely constrained within the dies, meaning the metal is fully enclosed by the dies, and there is no chance for excess metal to form a flash. This is why it's also known as flash less forging. For successful closed die forging, the precise amount of material is crucial. If too much metal is used, it can damage the dies since there is no space for the excess to escape. Conversely, if too little metal is used, it won't completely fill the die cavity, resulting in an incomplete part. Therefore, accurate calculation and measurement of the metal volume are essential to ensure the die cavity is properly filled without causing damage.

Impression Die Forging

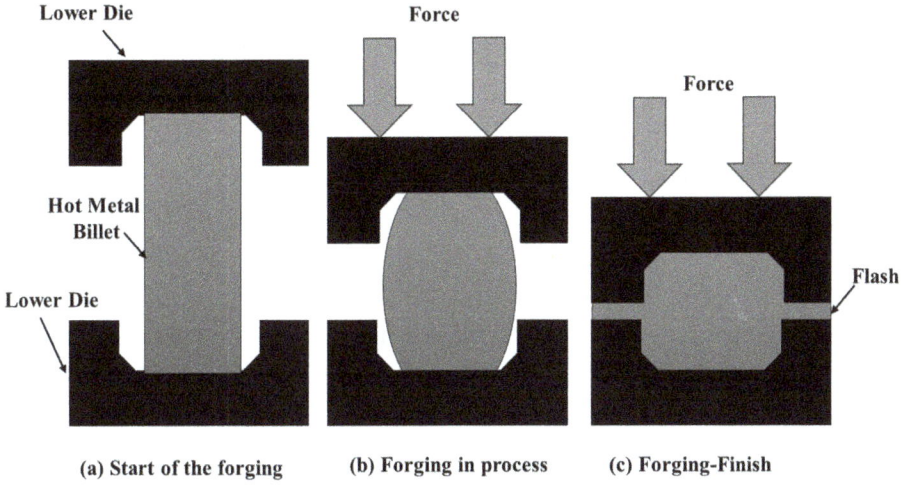

(a) Start of the forging   **(b) Forging in process**   **(c) Forging-Finish**

Figure 2.4: Impression die forging.

Close Die Forging

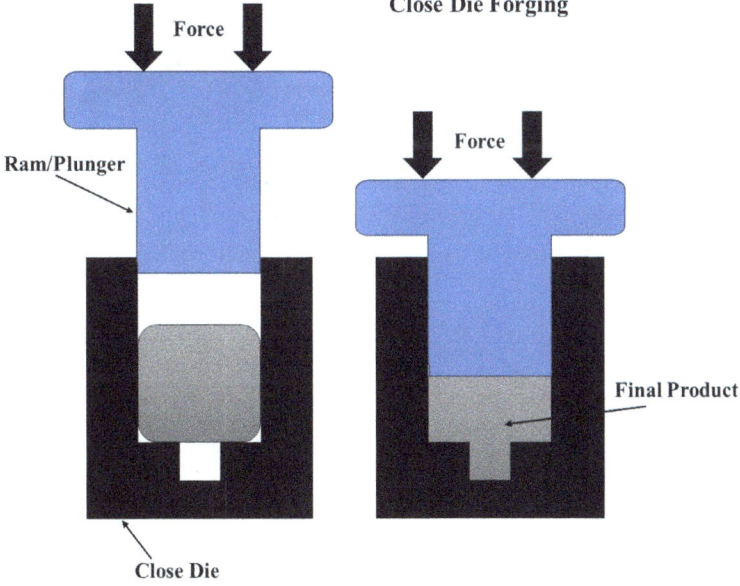

Figure 2.5: Closed die forging.

4. Upset forgings

Upset forging is performed at the end of the rod to increase the cross-sectional area of the rod for a particular length as in bolts, screws, etc. This process involves longitudinal compression of the rod. Usually upsetting are two types (see Figure 2.6):

(a) Open upset forging

(b) Closed upset forging

If the length to diameter ratio of rod is too high, there may be high chances of buckling of the rod.

5. Roll forging

Roll forging is a metalworking process where a heated metal billet is passed through a set of cylindrical rolls with grooves, which shape the metal into the desired cross-sectional form. Roll forging, shown in Figure 2.6(c), is often used to produce long, uni-

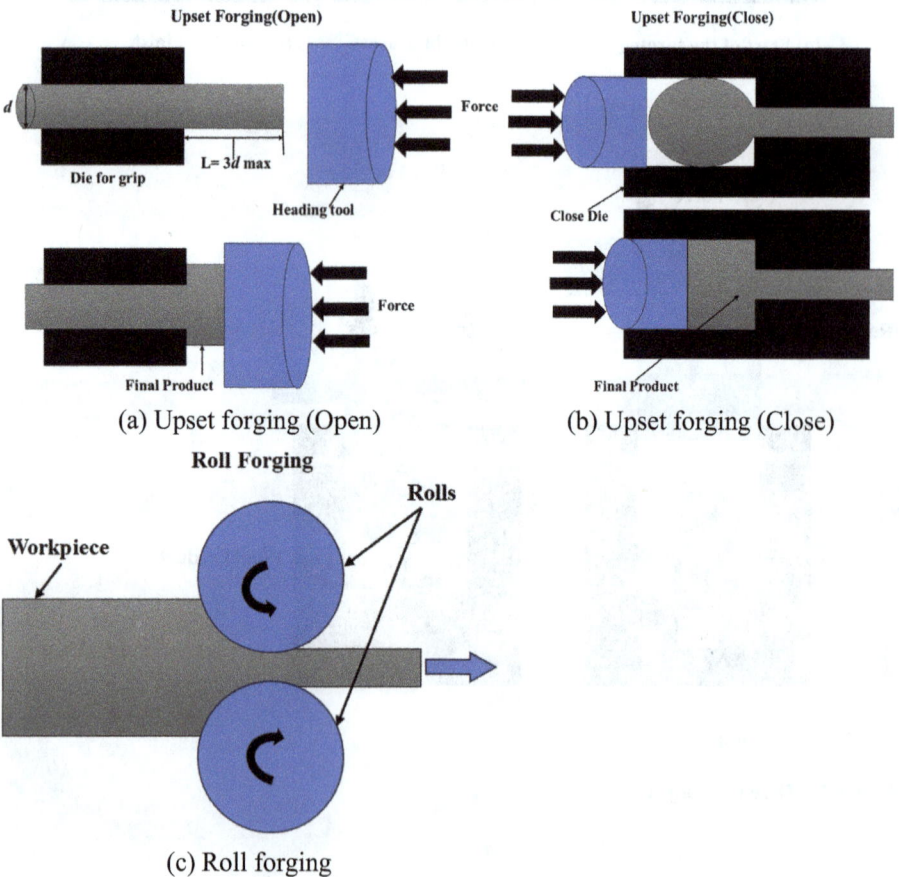

Figure 2.6: (a) Upset forging – open; (b) upset forging – closed; (c) roll forging.

form products such as axles, tapered levers, and leaf springs. Overall, roll forging is a versatile and efficient method for producing a variety of metal components with improved mechanical properties and precise dimensions.

6. Swaging

This operation involves hammering hot metal from various angles, causing the metal to flow inside the die to achieve the desired shape. In this process, the die itself functions as a hammer, repeatedly striking the metal to form it. This method is particularly used to reduce the diameter of rods and tubes. The process can be performed manually or with the assistance of power hammers and presses, ensuring that the metal is evenly shaped and sized.

By working the metal from different angles, the operation ensures uniformity and precision in the final product. The repeated hammering not only shapes the metal but also enhances its mechanical properties by refining the grain structure and increasing its strength. This technique is often employed in manufacturing components that require high strength and durability such as automotive parts, aerospace components, and heavy machinery elements.

2.2.2 Forging operations

1. Drawing out: In this operation force is applied in a direction perpendicular to the longitudinal axis. This operation is used to reduce the diameter and increase the length of workpiece.
2. Upsetting: This operation is opposite to the drawing out in which cross-sectional area increases and length of stock reduces. In upsetting force is applied in a direction parallel to the longitudinal axis to reduce the length and increase the area of cross-section.
3. Heading: In this operation only a portion of the job need to be forged, e.g., forging hand of bolts, valves, coupling, rivets nails, etc.
4. Fullering: It is the operation of reducing the thickness of metal at middle and increase the length of stock (Figure 2.7(a)).
5. Edging: In this type of operation metal is displaced from edge to inside. This process uses to cup shape dies which create bulge in middle (Figure 2.7(b)).
6. Bending: In bending operation, the metal is stressed in both tension and compression at values below the ultimate of the material without appreciable change in Figure 2.7(c).

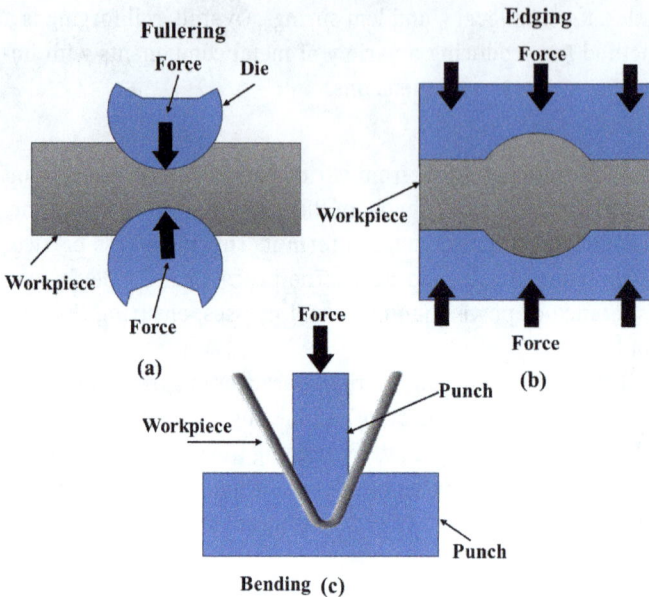

Figure 2.7: (a) Fullering, (b) edging, and (c) bending operation.

2.2.3 Forging defects

Common defects, as shown in Figure 2.8(a)–(h), occur due to improper working temperature, die design, cooling rate, etc., some defects are discussed below:

a) Cold shut: This defect is caused by poor die design or due to incorrect positioning of workpiece in the die cavity.

b) Die shift/mismatch: This defect is caused by misalignment of upper and lower die.

c) Pitting: After the forging operation, die must be clean; otherwise, scale is forced to the surface of workpiece during forging in the same die. This defect is called as pitting.

d) Incomplete die filling: When proper amount of material is not selected, insufficient number of hammer blows and low temperature of stock (metal to be forged) and wrong die design: metal flow in dies will not takes place properly and some portion of cavity will remain vacant.

e) Fins and rags: Small projection or excessive metal rejected from the die surface remain attached to the forging. These projections are called fins and rags.

f) Surface cracks: Cracks are formed due to various reasons such as rapid cooling of forged material, improper heating, and low forging temperature. Longitudinal and transverse cracks are formed in forging.

g) Flakes: Flakes are the internal break and ruptures. These cracks are formed due to rapid cooling of forged product.

h) Decarburization: When material is subjected to very high temperature for a long period of time, carbon is lost due to oxidation, and this phenomenon is termed as decarburization.

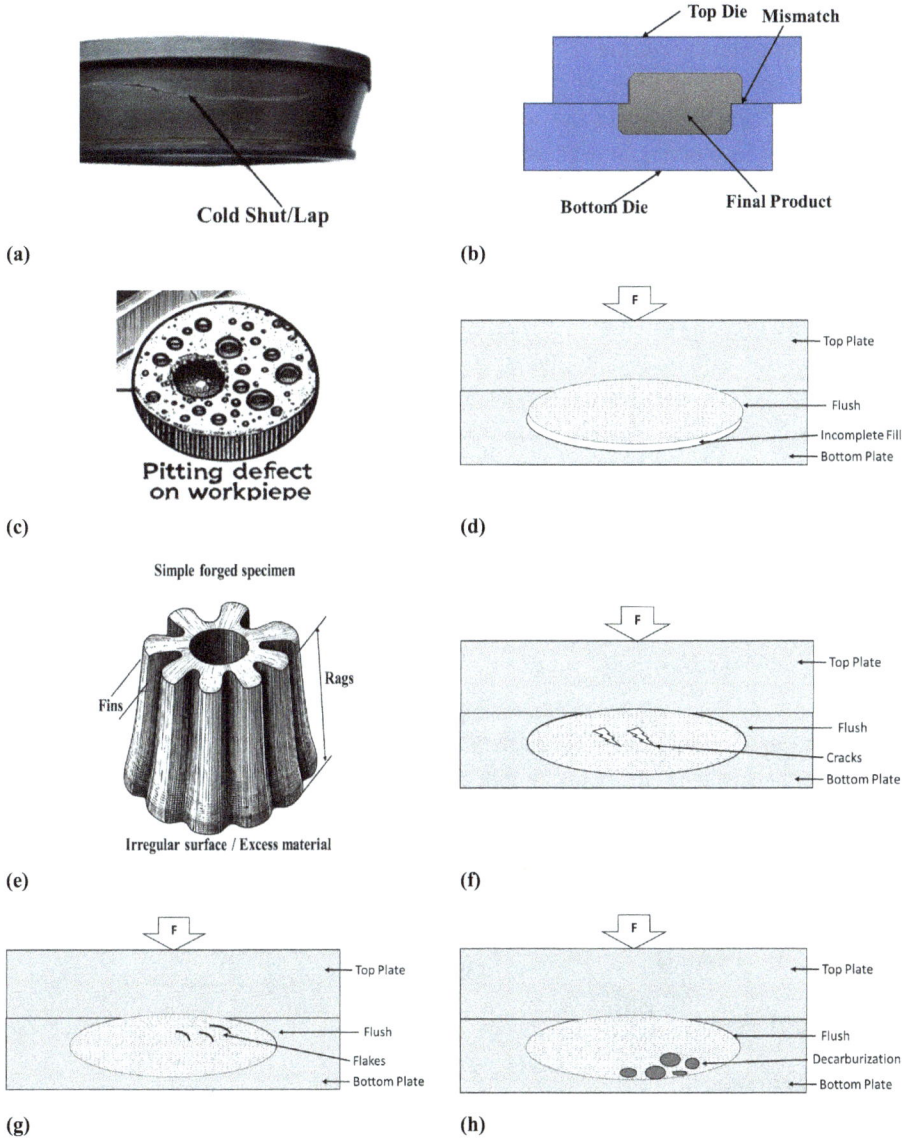

Figure 2.8: Common defects in forging: (a) cold shut/lap; (b) mismatch between dies; (c) pitting defect; (d) incomplete fill; (e) fins and rags; (f) surface cracks; (g) flakes; (h) decarburization.

2.2.4 Rolling operation

Rolling is a metal forming process in which metal undergoes plastic deformation as it passes between rotating rolls. This process is primarily used to reduce the thickness or cross-sectional area of a workpiece by applying compressive forces through a set of rolls. Rolling can be performed either above or below the recrystallization temperature.

The process occurs in two stages: first, the cast metal ingot is transformed into blooms, billets, and slabs. In the second stage, these blooms and slabs undergo hot rolling to produce plates, sheets, and other forms. To enhance surface finish and dimensional accuracy, some slabs may also undergo cold rolling.

1. **Mechanics of rolling operation:** The fundamental rolling process is schematically illustrated in Figure 2.9. In this process, a strip with an initial thickness h_i enters the roll gap and is compressed by the rotating rolls, reducing its thickness to h_t. The rolls apply a compressive force, and their surface moves at a velocity V_r.

To maintain a constant volume flow of metal, the strip's velocity increases as it exits the roll gap. While V_r remains uniform along the roll gap, the difference in strip velocity at the entry and exit results in relative sliding between the roll and the strip.

At a specific point within the contact zone, the roll's surface velocity equals the strip's velocity. This is known as the neutral point or no-slip point. To the left of this point, the roll velocity (V_r) is greater than the strip's entry velocity (V_i), causing the roll to grip the strip. To the right of the neutral point, the strip's exit velocity (V_f) exceeds the roll velocity (V_r), leading to relative motion between the strip and the roll.

Let:
- h_i and h_f be the thicknesses of the strip before and after rolling;
- V_i and V_f be the velocities of the strip before and after rolling;
- V_r be the surface velocity of the roll;
- b_i and b_f be the widths of the strip before and after rolling.

Assuming the width remains constant ($b_i = b_f = b$), the volume flow rate must be conserved. Thus, the input volume equals the output volume:

$$V_i \, h_i \, b = V_f h_f b$$

This simplifies to $V_f = \dfrac{V_i h_i}{h_f}$

2. **Types of rolling**
a) Two high non-reversing rolling mills/stand
b) Two high reversing rolling mills/stand
c) Three high rolling mills/stand
d) Four high rolling mills/stand
e) Cluster mills

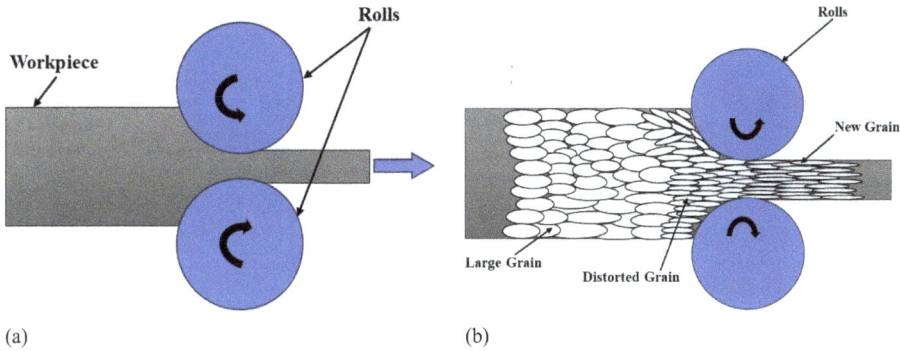

Figure 2.9: (a) Mechanics of rolling process with geometry; (b) rolling process showing changes in grain structure.

a) **Two high non-reversing rolling mills/stand:** In this arrangement two rolls are used as shown in Figure 2.10(a). These two rolls are supported in bearing and move in one direction only. The upper roll may be move up and down direction to adjust the distance between the rolls in order to get desired thickness of sheet.

b) **Two high reversing rolling mills/stand:** In this arrangement, direction of rotation of both rolls can be reversed as shown in Figure 2.10(b). This rolling reduces the handling of hot metal during rolling passes, but there is more consumption of power than non-reversing rolling stand.

c) **Three high rolling mills/stand:** In this arrangement three high rolling stands are used. Directions of rolls and workpiece are shown in Figure 2.10(c). This system increases the production rate.

d) **Four high rolling mills/stand:** In this arrangement shown in Figure 2.10(d) there are two working rolls and two backup rolls. Backup rolls provide the stability and rigidity to the working roll which are small. Backup rolls also prevent deflection of working rolls.

e) **Cluster mills:** Cluster mills, represented in Figure 2.10(e), are used to roll foils. It has two working rolls which are smaller in diameter and four or more back up rolls. These backup rolls provide support to working rolls.

2.2.5 Cold rolling

Cold rolling is employed in the production of sheets, strips, bars, and rods when smooth surfaces and precise dimensions are required for the final products. This process is typically conducted on four-high and cluster-roll mills. Cold rolled sheets and strips are available in four variations: skin-rolled, quarter-hard, half-hard, and full-hard:

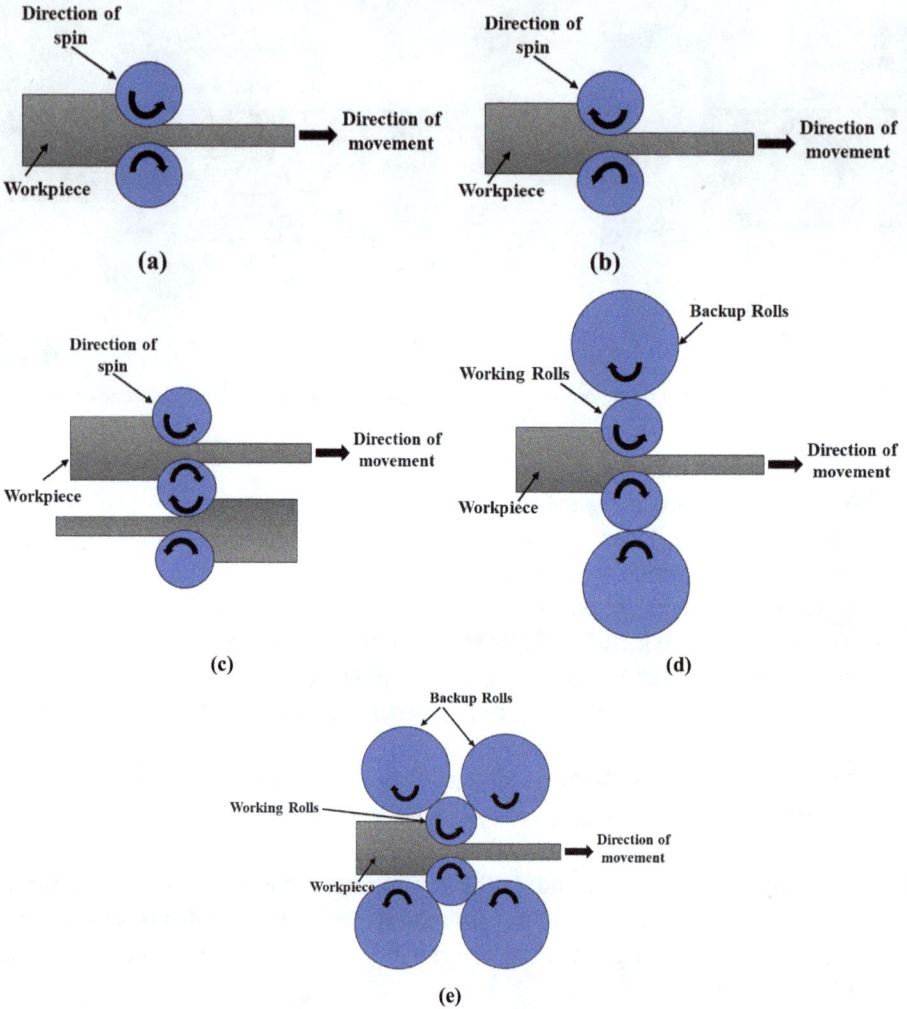

Figure 2.10: (a) Two high non-reversing rolling mills; (b) two high reversing rolling mills; (c) three high reversible rolling mills; (d) four high rolling mills; (e) cluster mills.

1. Skin-rolled metal produces smooth surface and uniform thickness on 0.5–1% reduction. It can be further cold-worked.
2. Quarter hard, half hard, and full hard undergo further cold reduction up to 50%; thus their yield strength increase and ductility decrease. They have strong directional properties. Quarter-hard plates can be bent without breaking across the grains, half hard by 90°, and full hard by 45°, respectively, about a radius equal to its thickness.

– **Rolling defects:** Some common defects that occur during rolling operation are pits, crack, rust, scale, etc. These defects are due to impurities in metal and inclusions. Apart from these defects there are some more defects shown in Figure 2.11, which are discussed below:
   a) Wavy edges: Wavy edge is formed on work piece due roll bending.
   b) Alligatoring: This usually occurs in rolling of slabs especially of aluminum and molybdenum alloys where the work piece splits along a horizontal plane at the exit. The top and bottom parts follow the rotation of their respective rolls.
   c) Edge cracking: Due to limited ductility of work material or uneven deformation especially at edges, edge cracks occur during rolling of ingots, slab, and plates.
   d) Folds: Folds are the defects that occur during rolling of ingots, slabs, and plates.
   e) Centerline crack: This is the crack formed on the center of work piece.

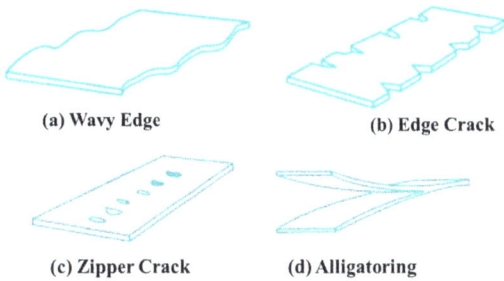

(a) Wavy Edge   (b) Edge Crack

(c) Zipper Crack   (d) Alligatoring

Figure 2.11: Rolling defects.

2.2.6 Extrusion process, products, and applications

Extrusion is the process of squeezing a metal through shaped die to get product of uniform cross-sectional area as shown in Figure 2.12. This process is used to produce uniform cross-sectional profile. Extrusion has some basic characteristics: mechanical properties are superior to rolling, it gives good surface finish, and complex shape can be easily extruded. It has certain disadvantages:
– Short tool life due to high constant stress;
– Productivity inferior to rolling;
– Structural inhomogeneity;
– Different flow rate of external and internal layers;
– High tooling cost;
– Hollow shape are not recommended and extremely thin section should be avoided.

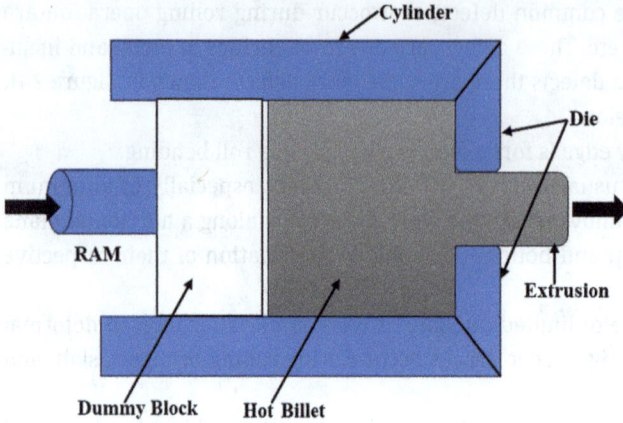

Figure 2.12: Extrusion process.

- **Application:** Building construction, household appliances, milk cartons, PVC pipes, torch body, etc.
- Classification of extrusion processes as depicted in Figure 2.13 is given below.

1. Hot extrusion
   i.   Forward extrusion
   ii. Backward extrusion
2. Cold extrusion
   i.   Impact extrusion
   ii. Hydraulic extrusion

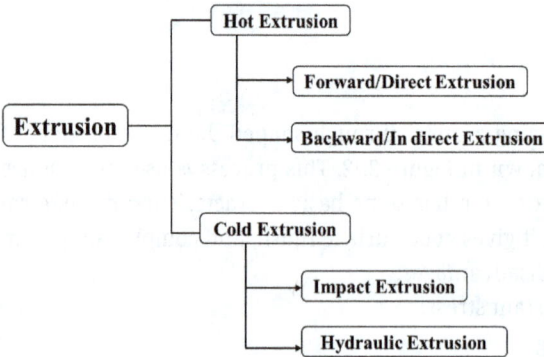

Figure 2.13: Types of extrusion process.

The extrusion processes are now briefly discussed in the following paragraphs.

1. Hot extrusion

Hot extrusion is carried out above the recrystallization temperature. The working temperature of between 55% and 75% of melting temperature and pressure range is from 40 to 700 MPa. It has certain advantages and limitations. Extrusion forces can be drastically reduced, and brittle material can also be extruded using backward extrusion, low tooling cost, and large deformation possible.

- **Limitation**
    - Formation of oxide on outer skin of billet affecting the flow pattern
    - Poor surface finish
    - Due to higher temperature, special equipment needed
    - Die wear rate are high

2. Cold extrusion

Cold extrusion is carried out below the recrystallization temperature of metal being extruded. Extrusion pressure is higher than the hot extrusion. It gives the better surface finish and improves mechanical properties.

- **Direct extrusion/forward extrusion**

In this extrusion billet is pushed in a container through a die as shown in Figure 2.14. The direction of movement of ram and extruded product are in the same direction. It is the most commonly used extrusion process. The operation consists of placing the billet in container. Container is heavily walled and contains Ram and die. Ram is used to push the billet through die.

- **Indirect-extrusion/backward extrusion**

In this type of indirect extrusion, as shown in Figure 2.14, the directions of extruded product and force are in opposite direction. In this process die is stationary and the billet and container move together. In order to keep the die stationary, a ram is used which should be longer than container. Since the billet move with the container, therefore, there will not be friction between billet and container. So maximum friction can be reduced with this type of arrangement. This has following advantages:

i. Up to 30% reduction in friction
ii. The extrusion have low tendency to crack as there is no friction heat
iii. Life of container liner is longer as there is less wear.

i. Impact extrusion

It is also known as reverse (backward) cold extrusion and is shown in Figure 2.15(a). The metal blank or slug for the making the component is placed in the die and the punch strikes the slug against the die. This impact on slug causes the slug material to get extruded through the gap between the punch and the die in the direction opposite

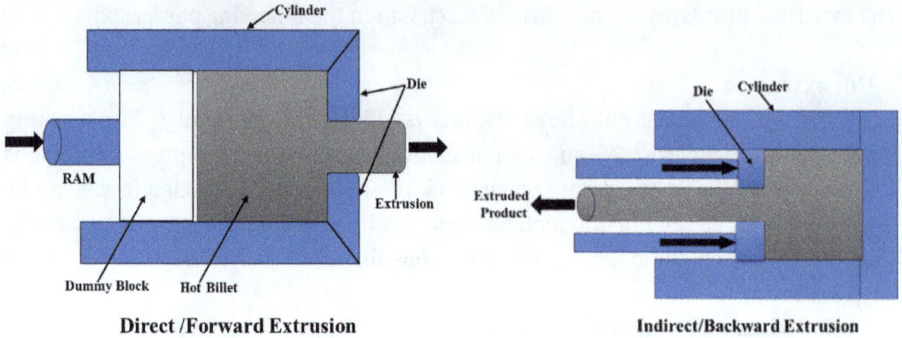

Figure 2.14: Types of direct and indirect extrusion.

to the punch movement. The sidewalls go straight along the punch. The process is used for making collapsible tubes for toothpaste medicine toothpaste, torch body, shell cases, soft drink canes, etc. Advantages of the process include

– High processing speed
– Less wastage of metal blank
– Good dimensional tolerance

ii. Hydrostatic extrusion

In this extrusion billet is completely surrounded by pressurized fluid. It reduces or eliminates friction on container walls. This extrusion squeezes the metal billet by applying uniform pressure on it. Thus higher deformation is achieved. Hydraulic fluid acts as a lubricant at billet die interface (sealing should be proper to contain the leakage). The uniform pressure compresses the metal billet from all sides eliminating the formation of cracks; hence the process is most suited for extruding brittle materials. The fluid pressure range between 1,100 and 3,100 MPa. Fluid used are castor oil, ethyl glycol, etc. The setup is shown in Figure 2.15(b).

2.2.7 Wire and tube drawing

It is the operation of pulling the solid rod, wire, or tube through a die in order to reduce its cross-sectional area. The operation consists of pulling of pointed, tapered wire, or tube through a conical die to reduce its thickness. Die is made up of hard material. The wire will take shape of die opening. This operation improves strength and hardness due to increase in density of reduced cross-section:

i. Wire drawing
ii. Tube drawing

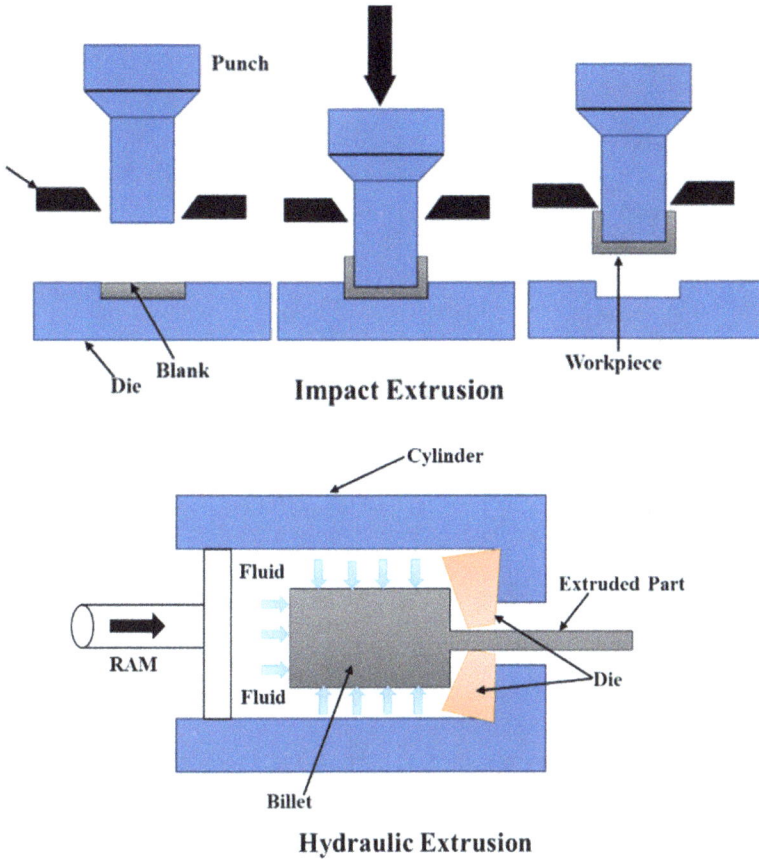

Figure 2.15: (a) Impact extrusion; (b) hydrostatic extrusion.

i. Wire drawing

It is the operation of reducing the cross-sectional area of wire or solid rod. In this operation wire or rod is being pulled through a die. Since volume of metal remains the same at entry and exit of die, the length of the wire depends upon new cross-section of wire. When wire or rod is pulled through a die there will be three forces on metal inside the die. Wire drawing sectional view is shown in Figure 2.16:

–   Tensile force by which rod is pulled
–   Compressive force due to die pressure
–   Friction force due to friction between die and material

Due to friction, heat will generate that helps to reduce draw force while getting plastic deformation in material.

Wire Drawing

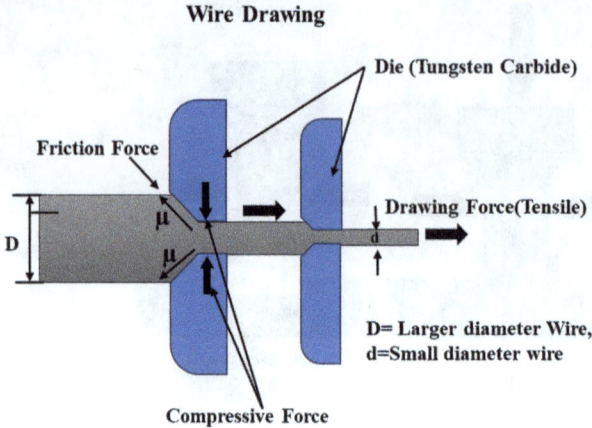

Figure 2.16: Wire drawing sectional view.

— **Process characteristics**
a) In a wire drawing operation, a rod is pulled through a conical die to reduce its diameter.
b) As the diameter of the wire decreases, its length increases.
c) A series of dies with progressively smaller diameters may be used to achieve very small diameter wires.
d) Certain mechanical properties, such as hardness and strength, are improved through the process.
e) The process requires careful control of the drawing speed and lubrication to prevent wire breakage.
f) Wire drawing can be applied to various materials including metals like copper, steel, and aluminum.

ii. Tube drawing

In the tube drawing process as shown in Figure 2.17, a hollow tube is pulled through a die with the assistance of a mandrel or plug, which supports the tube's interior and helps control its inner diameter. The primary function of the plug is to facilitate wall thickness reduction and maintain dimensional accuracy. However, if controlling the inner diameter or reducing the wall thickness is unnecessary, the mandrel can be omitted. In such cases, the process is known as tube sinking as shown in figure 2.17 (a).

When drawing a tube over a stationary mandrel, the maximum feasible cross-sectional area reduction is typically limited to 40% per pass due to the increased friction between the tube and the mandrel. However, a reduction of up to 45% can be achieved if a floating mandrel is carefully designed to move within the die throat, sometimes called as moving mandrel, as shown in figure 2.17 (d). In addition to allowing greater material reduction, floating mandrel drawing requires lower drawing forces compared to the use of a fixed mandrel as shown in figure 2.17 (c).

It is important to note that in this process, tool design and lubrication play a crucial role, as achieving proper alignment and minimizing friction are particularly challenging in floating mandrel drawing refer figure 2.17 (b).

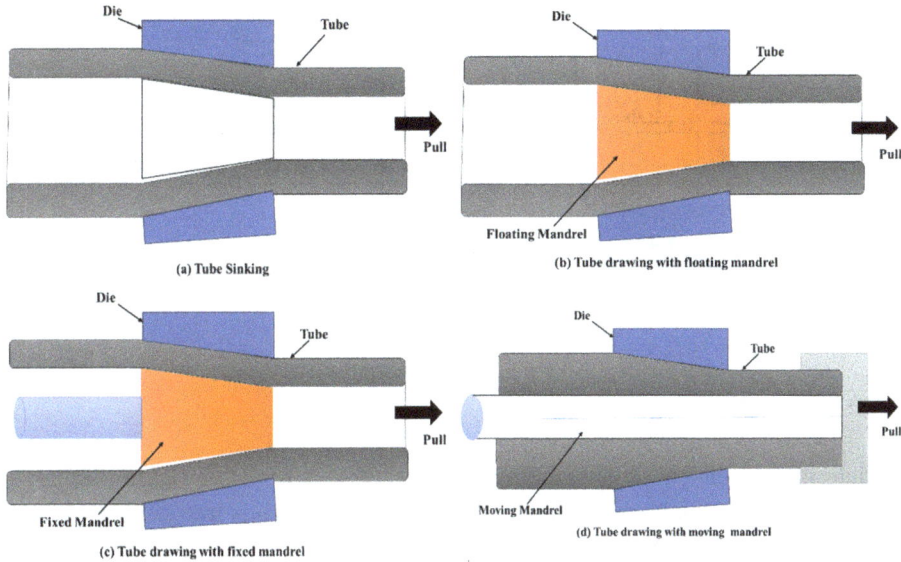

(a) Tube Sinking

(b) Tube drawing with floating mandrel

(c) Tube drawing with fixed mandrel

(d) Tube drawing with moving mandrel

Figure 2.17: Tube drawing processes: (a) sinking; (b) floating; (c) fixed mandrel; (d) moving mandrel.

2.3 Basic press working operation

Sheet metal working involves the cold working of the metals generally in the form of thin sheets. This process is carried out at room temperature that leads to get higher dimensional accuracy, close tolerance, and good surface finish on the worked parts. After press working, no machining is required. This type of operation is generally applied for mass production. The force needed to cut the workpiece is provided by a machine called press machine.

2.3.1 Die and punch assembly

A simple die and punch assembly is fitted on press machine as shown in Figure 2.18. Also called die-set. The parts of the die and punch system are labeled. On this set of die-punch, various operations are performed, such as cutting, drawing, forming, and bending.

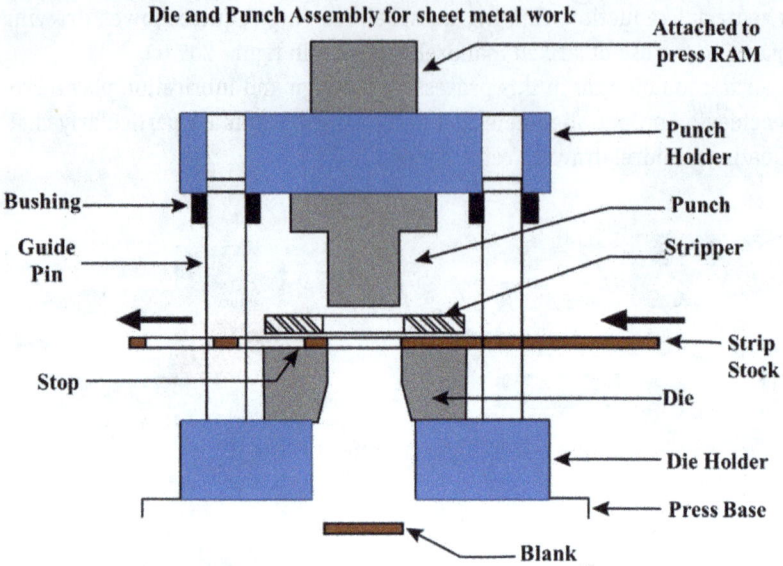

Figure 2.18: Die and punch assembly.

Various parts of die-punch assembly are now discussed in brief (see Figure 2.18).

a) Press ram (slide): Press has a punch at its bottom, which is free to move up and down in order to perform various operations.

b) Shank of punch: Shank is the upper part of punch holder. With the help shank, punch holder holds the tool called punch.

c) Punch: Punch is nothing but a tool. It is a male member of the unit (die-punch set). It is made up of wear resistant and hard steel. It has good strength to pierce in metal.

d) Punch holder (upper shoe): It holds the punch and is attached to the lower end of the press ram. It is made up of cast steel.

e) Bushing for guide post: This provides proper alignment to the punch and die and also it holds the same. During press working proper alignment must be there between die and punch otherwise product will be defective.

f) Stripper plate: When the punch moves up after work, stripper plate makes the punch free from the metal sheet.

g) Die-holder (lower shoe): It is a female member of the unit. It contains the die cavity.

h) Backup plate: This plate provides backup to the punch holder.

i) Bolster plate: This is the base to provide support to the die.

2.3.2 Cutting process (sheet metal working/press working/shearing operation)

Cutting is a fundamental operation in sheet metal working, where a workpiece is separated into desired shapes using applied force. It is typically performed using press working machines and involves shearing, blanking, punching, and other processes. A brief introduction of some important sheet-metal working operation is given below:

a) **Blanking:** Blanking is the press working operation in which a particular size part is cut from the metal strip. The cut part from the metal strip is called blank and is used for further working. The reaming metal strip is called scrap. Structure of blanking has been shown in Figure 2.19(a). Example: For production of coin, first of all a circular disc (coin shape) of metal is cut from the metal strip, and these discs are now further be worked to impose the impression.

b) **Punching:** It is an operation of cutting the metal to produce hole. In this operation slug is cut from the metal strip and this slug is scrap (waste). It is the reverse of blanking operation. Punching has been shown in Figure 2.19(b).

c) **Shaving:** Shaving is the operation in which surface is finished by cutting rough edges of workpiece which is previously cut by some other operation. This operation is done to finish, smooth and brought within close dimensional tolerance (refer Figure 2.19(c)).

d) **Trimming:** When the workpiece is held to work on press machine, the gripped area become marked and form flashes. Trimming, as shown in Figure 2.19(d), is done to remove these marks and flashes from the periphery of the workpiece which is previously formed by some other process.

e) **Notching:** Notching is a cutting operation of very small indentation in the edge of the metal strip. It has been represented in Figure 2.19(e).

f) **Slitting:** Slitting refers to the operation of making incomplete holes from s slit (an opening) in the sheet metal as shown in Figure 2.19(f).

g) **Lancing:** Lancing, shown in Figure 2.19(g), is cutting along a line in workpiece without removing the slug from the workpiece. In this operation hole is partially punched and one side bend down as a louver.

h) **Nibbling:** It is an operation of cutting the contour part/narrow slots/curved shape from the metal strip. A small piece of metal is removed from metal strip with each stroke of the punch. It produces a cut of negligible width. Figure 2.19(h) shows the structure of nibbling process.

i) **Embossing:** Embossing is a process which produces shallow indentation or raised designs and there in no change in metal thickness. Embossing means producing shapes as letters or fancy design on thin materials. In this type of workpiece design shows in Figure 2.19(i) on both sides. Complete detail on one side and relief detail on other side is made in embossing.

j) **Coining:** Coining is a kind of cold forging in which all surfaces of the work are restrained. In the coining process, first design is made on punch and die. Then blank (metal sheet) is placed in the die cavity. When the punch fall on the blank with heavy pressure, and metal is made to flow into the cavities (design) made on punch and die until the design is reproduced on the blank. Shallow configuration on the surface of flat object like coin is made by this process shown in Figure 2.19(j).

k) **Hobbing**: Hobbing, shown in Figure 2.19(k), is the operation of producing mold cavity in soft steel by using die.

l) **Shearing:** It is also known as die-cutting process shown in Figure 2.19(l). It is the process of cutting sheet metal without forming chips or melting the material. It involves two sharp-edged tools (upper and lower blade) that apply force to cut the material. Common examples include guillotine shearing and slitting.

m) **Perforating:** Perforating is the operation of making many identical holes and continuous rows or uniformly arranged. It is shown in Figure 2.19(m).

n) **Parting:** Parting is an operation of cutting a sheet into two or more pieces as shown in Figure 2.19(m).

o) **Spinning:** Spinning is the operation to obtain seamless, hollow, circular, or cylindrical shapes. In this process a rotating disc of metal strip is formed over a male form of metal by applying localized pressure to the outside of the disc with the help of simple, round-ended metal or by roller. This operation is applied only to thin material. Figure 2.19(n) shows the basic spinning process.

–   **Factors affecting cutting performance:**
    –   Material type: The hardness and thickness of the sheet affect the cutting force required.
    –   Clearance: The gap between the punch and die affects the cut quality and tool life.
    –   Tooling: Sharpness, material, and alignment of the punch and die determine accuracy.
    –   Cutting speed: Higher speeds may improve efficiency but can cause tool wear.

–   **Other processes used in sheet metal are**
i. **Deep drawing**: Deep drawing is a metal forming process used to create cup-shaped components. It involves shaping a flat metal sheet into a cylindrical or box-like structure by pressing it into a die cavity with a punch. This technique is also effective for producing shallow parts and is widely applied in manufacturing household cookware, food containers, automobile fuel tanks, and various other hollow structures.

–   **Working principle:** As illustrated in Figure 2.20, the deep drawing process follows these steps:
a) A circular metal disc (blank) of diameter D and thickness t is placed over the die opening.
b) A blank holder applies a controlled force to keep the blank in position, preventing wrinkles during the process.

c) A punch with diameter d moves downward, pushing the blank into the die cavity.
d) As the punch descends, the metal flows radially and forms the cup shape while maintaining its thickness.

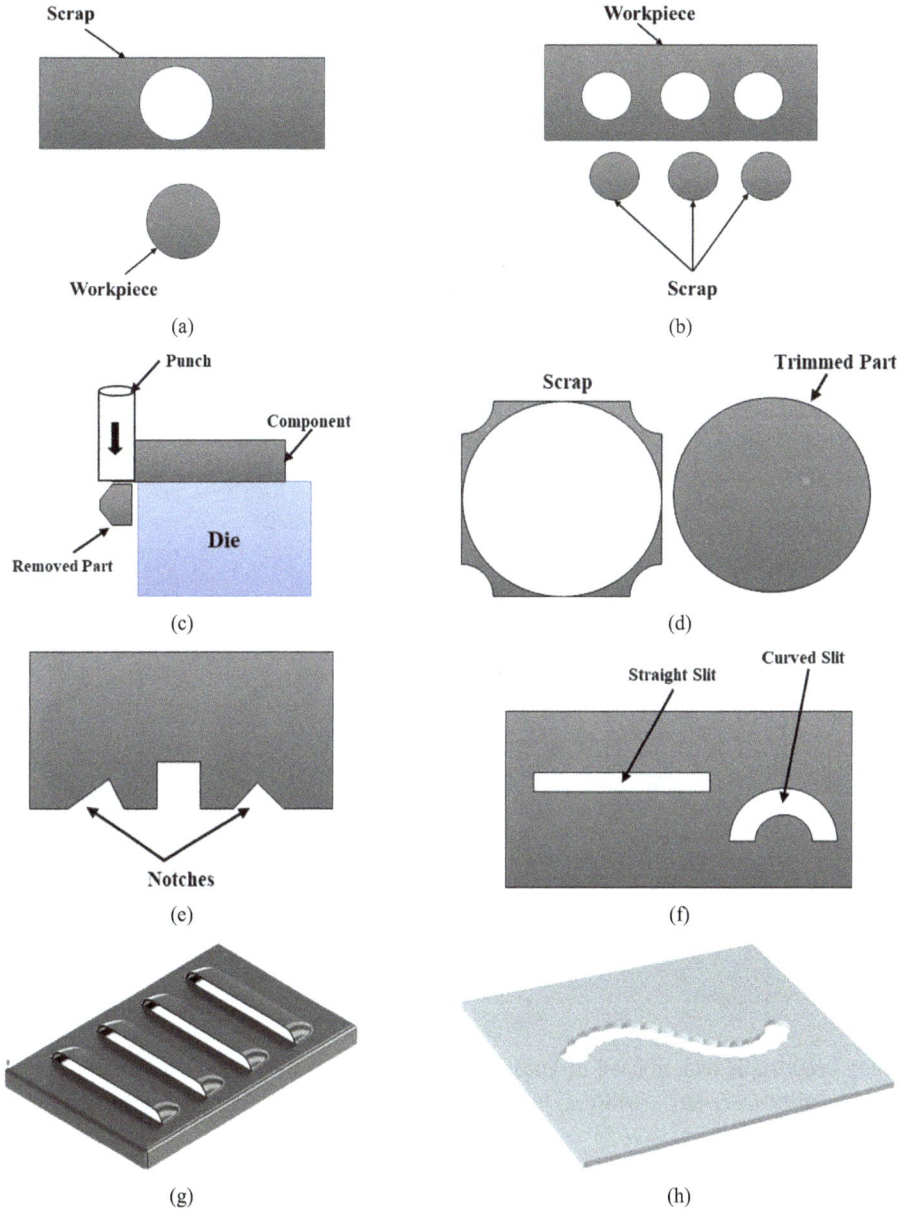

Figure 2.19: (a) Blanking, (b) Punching, (c) Shaving, (d) Trimming, (e) Notching, (f) Slitting, (g) Lancing, (h) Nibbling, (i) Embossing, (j) Coining, (k) Hobbing, (l) Shearing, (m) Perforating & Parting, (n) Spinning.

Embossing

Coining

(i)

(j)

Force (F)

(k)

(l)

(m)

(n)

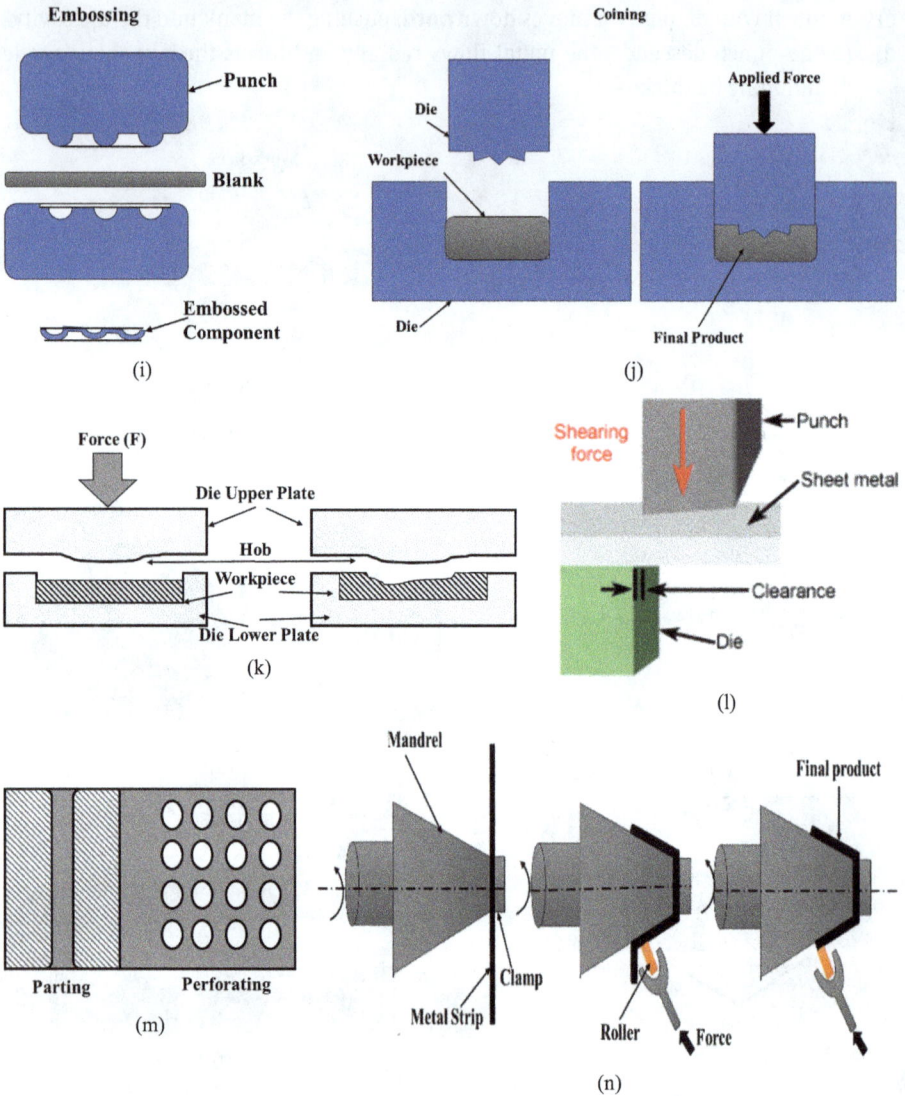

Figure 2.19 (continued)

Deep drawing is widely used in automobile, aerospace, and packaging industries due to its ability to create seamless, lightweight, and strong components with high precision.

ii. **Bending:** Bending, as illustrated in Figure 2.21, is one of the most common sheet metal forming operations. It is used to create features such as flanges, curls, and seams, while also imparting stiffness to the component.

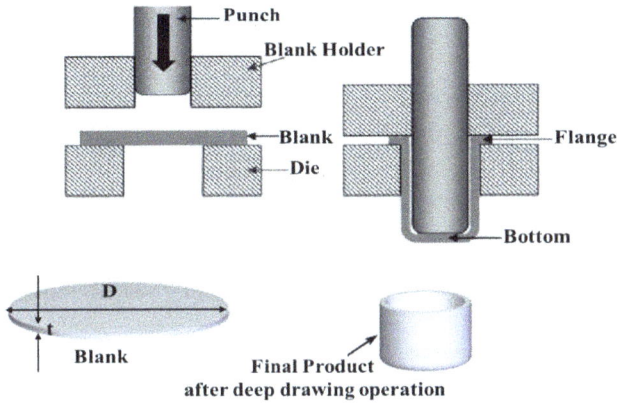

Figure 2.20: Deep drawing.

- **Working principle:**
  - During bending, the metal is subjected to both tensile and compressive stresses but remains below its ultimate strength, ensuring no significant change in thickness.
  - The outer fibers of the material are stretched (tension), while the inner fibers are compressed.
  - The neutral axis shifts toward the compression side, causing a larger portion of the material to experience tension.
  - On the tension side, the thickness of the material slightly decreases, and the width increases.
- **Springback effect**
  - After the bending force is removed, the material partially returns to its original shape due to its elastic recovery – this is known as springback.
  - To compensate for this, an additional over-bending is applied so that once the material springs back, it achieves the desired final shape.

This process is widely used in automotive, aerospace, and structural applications for forming precise and durable sheet metal components.

2.3.3 Application of sheet metal working

Sheet metal working is extensively used in both industrial and non-industrial applications, including

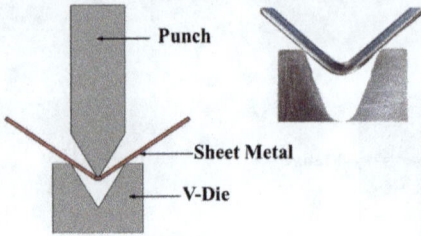

Figure 2.21: Bending operation.

i.   Aerospace industry
   – Aircraft fuselages, wings, body panels
   – Helicopter and spacecraft components
ii. Automobile industry
   – Body panels, bumpers, doors, chassis, and brackets
   – Exhaust systems and fuel tanks
iii. Construction industry
   – Roofing, home building, and structural components
   – HVAC ducts and metal frameworks
iv. Consumer goods and appliances
   – Kitchen equipment, office furniture, boilers
   – Refrigerators, washing machines, and microwave ovens
v.   Food and beverage industry
   – Cans, containers, and packaging materials
   – Food processing machinery

Due to its versatility, strength, and lightweight properties, sheet metal is widely used in manufacturing, infrastructure, and everyday consumer products.

2.3.4 Defects in sheet metal working

There are various types of defects can be seen in sheet metal formed products. Most of the defects are due to defective raw material, defective die design, and defective forming techniques. Defect due to raw material can be minimized by purchasing quality raw material and defect due to die design can be removed or minimized by proper design of die. The various defects caused due to processing/forming technique are now discussed below:
–   **Earing:** This defect occurs in deep drawing due to excessive or uneven drawing. To minimize or eliminate earing excessive deformation in deep drawing should be avoided.

2.4 Bench work and fitting

Bench work and fitting are essential processes used to complete and finish a job to the desired shape and size.
- Bench work refers to the manual production of components using hand tools on a workbench. It involves cutting, filing, drilling, and shaping materials to achieve the required form.
- Fitting involves assembling parts and removing excess material to achieve the proper fit. It is not always performed on a bench – it can be carried out at different locations depending on the application.
- Both processes are widely used in metalworking, machining, and repair work to ensure precision and proper assembly of components.

The main operations which are performed in bench and fitting work are classified as filing, sawing, chipping, dieing (external thread cutting), tapping (internal thread cutting), reaming, marking etc. All the operations are shown in Figure 2.22 (a) to (g).
a) Filing: A machining process using a file to remove small amounts of material for smoothening, shaping, or deburring metal or wood surfaces. It ensures precision and improved surface finish.
b) Sawing: Sawing is done to sever (cut)bar stock and shapes to suitable length. In some cases it has been used to produce desired shapes. For producing only a few parts sawing can be used to produce contoured shape (contoured sawing) where it turns out to be more economical.
c) Chipping: Chipping is the process of removing excess material from the finished surface by removing very small amount of material by chipping action using a straight chisel with rounded ends.
d) Thread cutting (die): Threads are of two types: (a) external thread (b) internal threads. External thread can be made on lathe, die, and stock (manual), with automatic die (turret), milling, and grinding. Similarly, internal threads can be made on a lathe, with tap and holder (manual, semiautomatic, automatic), automatic tap by milling. Now the threads are also made by rolling.
e) Tapping (thread tapping): The cutting of the internal threads using a multipoint tool is called thread tapping and the tool is called a tap. For cutting internal threads using a tap, a hole is made slightly larger than the minor diameter of the thread is made by drilling, boring or casting. Solid taps will form flutes, made from carbon or high speed steel, are used in a set of three, first two of tapered for alignment, then plug tap, and bottoming taps are used for finishing the threads.
f) Reaming: A finishing process using a reamer to enlarge or refine drilled holes, ensuring precise diameter and smooth surface. It improves hole accuracy for proper fitment of components.

g) Marking: A preparatory step in manufacturing where lines, symbols, or measurements are drawn on workpieces using scribers, punches, or markers, ensuring accurate cutting, drilling, or assembly.

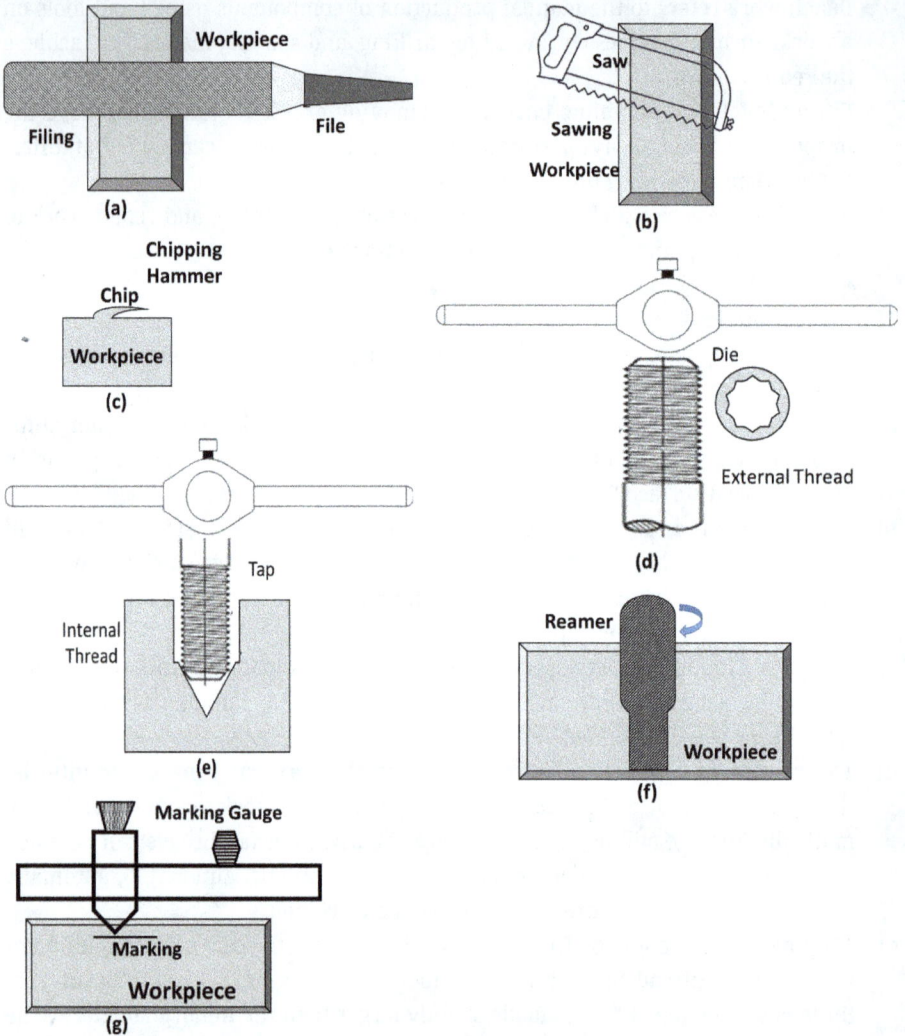

Figure 2.22: Types of operation in bench work: (a) filing; (b) sawing; (c) chipping; (d) dieing (external thread cutting); (e) tapping (internal thread cutting); (f) reaming; (g) marking.

2.4.1 Study of hand tool

There are several hand tools used in bench work and fitting some of them discussed below:

i. **Files:** Figure 2.23 illustrates a labeled diagram of a file, which is a hand tool used for cutting, smoothing, and fitting metal parts. They consist of a hardened and tempered blade with a pointed tang that is fitted into a wooden handle. Files cut only in the forward stroke and are classified based on size, tooth cut, grade, and shape.

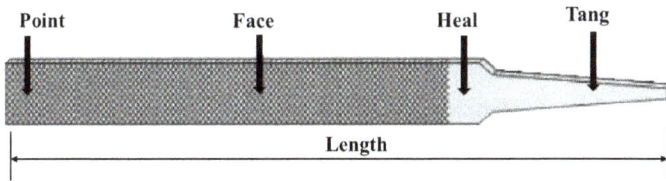

Parts of file.

– **Types of files:** Each type of file is designed for specific shaping and finishing tasks, making them essential tools in bench work and fitting operations. Its types have been shown in Figure 2.24 (a) to (f).
  a) **Flat file**
    – Tapered in width and thickness
    – Double-cut on the face, single-cut on the edges
    – Used for general-purpose filing on flat surfaces
  b) **Hand file**
    – Width remains parallel, thickness tapers toward the tip
    – Double-cut on both faces
    – Has one safe edge, making it useful where a flat file cannot be used
    – Used for finishing flat surfaces
  c) **Square file**
    – Square cross-section, double-cut on all sides
    – Tapered tip for fine adjustments
    – Used for filing square corners and enlarging square/rectangular slots
  d) **Round file (rat-tail file)**
    – Round cross-section, usually tapered (rat-tailed)
    – Parallel round files maintain a consistent diameter.
    – Used for filing curved surfaces, enlarging round holes, and forming fillets
    – Available in single-cut or double-cut

e) **Triangular file**
   - Tapered, double-cut, with a triangular cross-section.
   - Ideal for filing corners less than 90° and rectangular cuts.

f) **Half-round file**
   - Tapered, double-cut, with a cross-section of approximately one-third of a circle.
   - Used for filing curved surfaces and making round cuts.

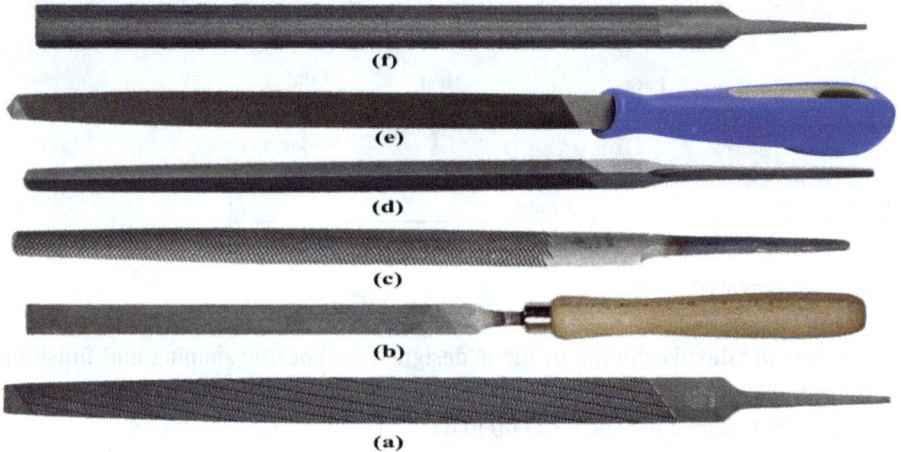

Figure 2.24: Types of files: (a) flat file; (b) hand file; (c) square file; (d) round file; (e) triangular file; (f) half round files.

ii. Vice types and their uses

A vice is one of the most commonly used tools for holding workpieces securely during bench work and fitting operations. Each type of vice, as shown in Figure 2.25 (a) to (e), serves a specific purpose, ensuring stability, precision, and safety in machining, metalworking, and assembly processes. Various types of vices are used depending on the applications:

a) **Bench ice**
   - Used for firmly holding workpieces.
   - Fixed to a bench using nuts and bolts.
   - Consists of a cast iron or steel body, square threaded screw, handle, and two jaws.
   - One jaw is stationary, while the other moves using the threaded screw.
   - Knurled jaw plates provide better gripping and can be replaced when worn out.

b) **Hand vice**
   - Designed for holding small objects such as screws, rivets, lock keys, and small drills.

- Typically 125–150 mm in length, with jaw widths of 40–44 mm.
- Used when bench vices are too large or inconvenient.

c) **Pipe vice**
- Specially designed for gripping round-section jobs like pipes and tubes.
- Grips at four points on the pipe's surface, ensuring a firm hold.

d) **Pin vice**
- Used for holding small round-section materials like thin rods, wires, and small files.
- Features a self-centering chuck mounted on a wooden or metal handle.
- The workpiece is secured by rotating the handle, tightening the chuck.

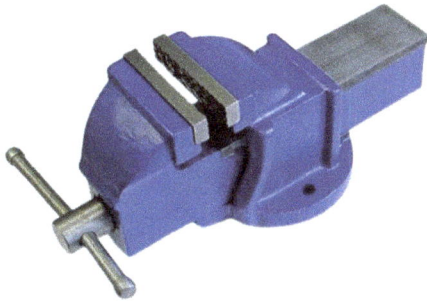 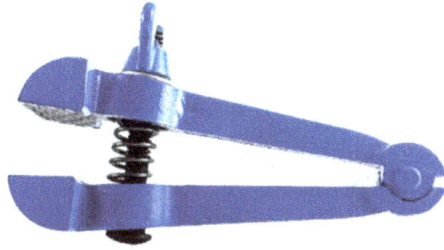

(a) Bench Vise   **(b) Hand Vise**

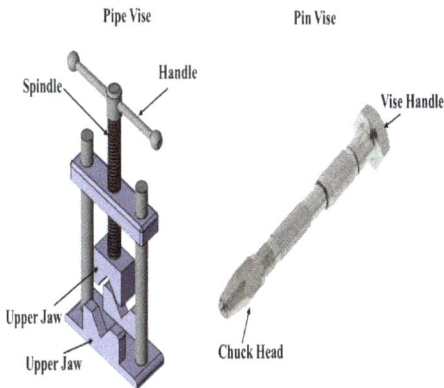 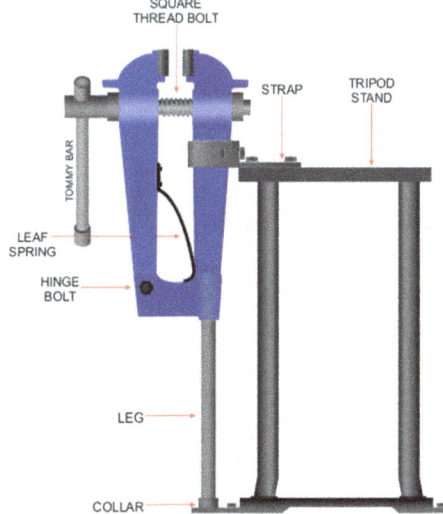

Figure 2.25: (a) Bench vice; (b) hand vice; (c) pipe vice; (d) pin vice; (e) leg vice.

e) **Leg vice**
   - Used in smithy shops and for heavy-duty tasks such as hammering, chipping, and cutting.
   - Secured to a workbench with an iron strap and bolted for stability.
   - Has a leg support that extends to the floor, absorbing impact forces.

iii. **Hammers, its types, and uses**

A hammer is a hand tool used for striking, shaping, bending, or assembling materials. It consists of a head (usually made of forged steel) and a handle (made of wood, fiberglass, or metal). Hammers come in various shapes and sizes, depending on their purpose. Parts of hammer have been described in Figure 2.26.

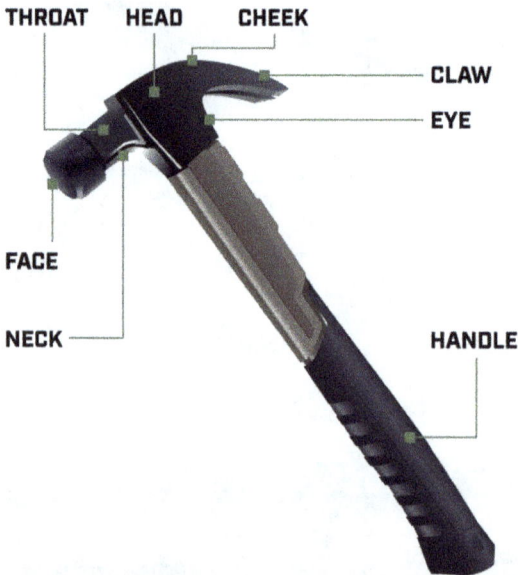

Figure 2.26: Parts of hammer.

Each hammer type has been shown in Figure 2.27 (a) to (g). They are designed for specific metalworking and fabrication tasks, ensuring precision and efficiency in different applications.

a) **Ball peen hammer**
   - Also known as a chipping hammer.
   - Has a rounded (ball-shaped) peen, which is hardened and polished.
   - Commonly used for riveting, shaping metal, and general striking tasks.

b) **Cross peen hammer**
   - Similar in size and shape to the ball peen hammer.
   - The peen is placed across the shaft (perpendicular to the handle).

    – Used for bending, stretching, and working on shoulders or inside curves.
c) **Straight peen hammer**
    – Has a peen parallel to the axis of the shaft.
    – Used for stretching or peening metal surfaces.
    – Useful in blacksmithing and sheet metal work.
d) **Soft hammer (mallet)**
    – Made of wood, rubber, or plastic.
    – Used when striking metal surfaces with minimal damage.
    – Ideal for woodworking, fitting delicate parts, and shaping sheet metal.
e) **Claw hammer**
    – Features a flat striking face and a curved claw for removing nails.
    – Commonly used in carpentry and construction.
f) **Sledgehammer**
    – A heavy-duty hammer with a large head for high-impact strikes.
    – Used in demolition, metal forging, and driving stakes.
g) **Club hammer (lump hammer)**
    – A short-handled heavy hammer with two flat faces.
    – Used for light demolition, masonry work, and driving chisels.

(a) Ball Peen   **(b) Cross peen** **(c) Straight Peen** **(d) Mallet or Soft** **(e) Claw**   **(f) Sledgehammer** **(g) Club or Lump**

Figure 2.27: Types of hammers: (a) ball peen hammer, (b) cross peen hammer, (c) straight peen hammer, (d) mallet, (e) claw hammer, (f) sledgehammer hammer, and (g) club or lump hammer.

iv. Chisel and its types

Chisel is used for removing metal by cutting and chipping away pieces of metal. Chisels are made of carbon steel. The cross-section of chisels is rectangular, hexagonal, or octagonal. They are manufactured by forging and hardened and tampered to achieve desirable characteristics. Types are represented in Figure 2.28 (a) to (g).

a) **Cross-cut chisel**
    – Designed for cutting grooves and keyways in shafts and wheels.
    – Used before applying a flat chisel for detailed shaping.
    – Has a narrow, straight cutting edge.

b) **Flat chisel**
   - Used for chipping and removing excess metal from a surface.
   - Has a slightly curved cutting edge for better efficiency.
   - Commonly used for general-purpose metal cutting.

c) **Half-round chisel**
   - Used for cutting oil grooves in bosses and pulley bearings.
   - Helps in creating pilot holes before drilling.
   - Features a rounded cutting edge.

d) **Diamond point chisel**
   - Ideal for cutting 'V' grooves, cleaning corners, and enlarging small holes.
   - Has a square-section tip ground at an angle to form a diamond shape.

e) **Side-cut chisel**
   - Used for removing excess metal from cotter ways and slots.
   - Generally operated by hand after drilling.

f) **Round nose chisel**
   - Used for cutting semicircular grooves and curved profiles.
   - Commonly applied in sheet metal work.

g) **Cape chisel**
   - Designed for cutting narrow grooves and slots.
   - Features a thin, flat cutting edge.

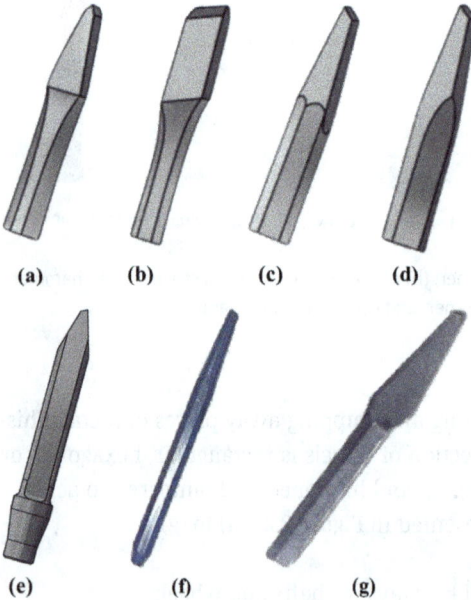

Figure 2.28: Types of chisels: (a) cross-cut chisel; (b) flat chisels; (c) half-round chisel; (d) diamond point chisel; (e) side-cut chisel; (f) round nose chisel; (g) cape chisel.

v. **Scraper and its types**

It is used to parting off thin slice or flakes of metal in order to get smooth surface. This is done by scraper which have hard cutting edge. Scrapers are shown in Figure 2.29 (a) to (e):

a) **Flat scraper**
   – Used for scraping flat surfaces to achieve a smooth and precise finish.
   – Commonly used in machine assembly and maintenance.

b) **Half-round scraper**
   – Designed for scraping curved or concave surfaces.
   – Often used in bearing housings and rounded machine parts.

c) **Triangular scraper**
   – Has three cutting edges for versatile use.
   – Suitable for scraping keyways, slots, and internal corners.

d) **Hook scraper**
   – Features a hooked cutting edge for cleaning internal bores and cylindrical surfaces.
   – Ideal for removing rust, scale, and old coatings.

e) **Bearing scraper**
   – Specially designed for scraping and fitting bearings to ensure proper contact.
   – Helps in achieving precision in bearing alignment.

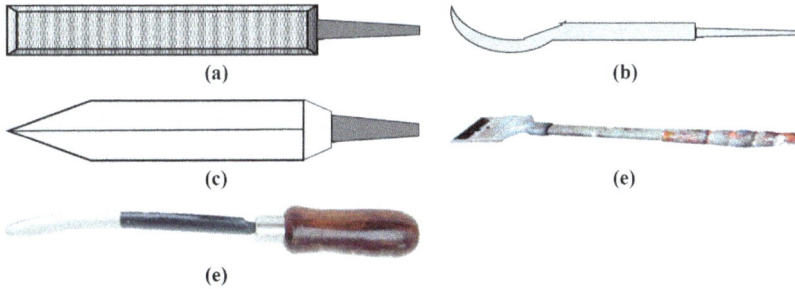

(a)   (b)

(c)   (e)

(e)

Figure 2.29: Types of scrapers: (a) flat scraper; (b) half-round scraper; (c) triangular scraper; (d) hook scraper; (e) bearing scraper.

vi. **Hacksaw**

Sawing is a process of removing the metal by hacksaw. It consists of a frame handle. A hacksaw is used for sawing all metals except hardened steel. A hacksaw, shown in Figure 2.30 (a) & (b), consists a frame handle and nut blade which is temporarily fixed by tightening screw. Frame holds the blade. Frames are two types:

a) **Solid frame**
b) **Adjustable frame**

In solid frame length cannot be changed but in adjustable frame we can change the length. They are made of high carbon steel, low alloy steel, or high-speed steel. Hacksaw is used to cut the harder metal as alloy steel.

Power hacksaw: It has driving mechanism with it. Drive is given by an enclosed motor. It is very similar to the hand hacksaw. Suitable mechanism is provided whereby length of the stroke and the weight applied can be varied.

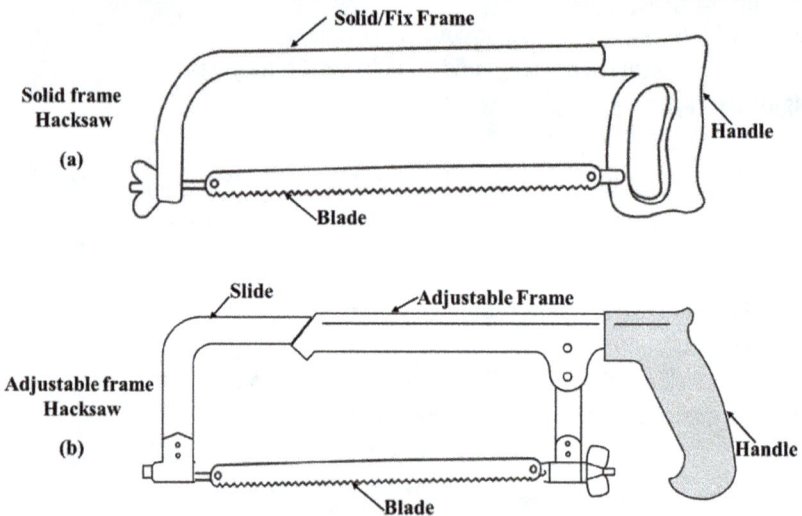

Figure 2.30: (a) Solid frame hacksaw; (b) adjustable frame hacksaw.

vii. **Marking tools**: In bench work and fitting the following tools are used for marking purpose. The various marking tools have been shown in Figure 2.31 (a) to (e).
   a) **Surface plate**: A surface plate is used for testing the flatness of a work piece and is also used for marking the work. Surface plate is used in small pieces of work while marking out table is used for larger jobs. Surface tables are made of gray cast iron and solid design or with ribs. They should be plane and re-flection-free illuminated and rest horizontally on a firm support, and the height of surface being about 800 mm from the floor.
   b) **Try square:** Try square is used to get two surfaces or edges at a right angle. It is made of a single piece. The sides and edges of any square piece may be checked by placing the beam of the square against the straight edge.
   c) **Vee block:** A vee (V)-block is made of steel with vee-shaped grooves. Round-shaped work pieces which are to be marked or drilled are firmly supported in a horizontal position and cannot be rotated easily.

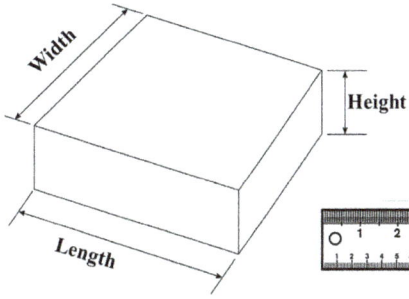

(a) Surface Plate

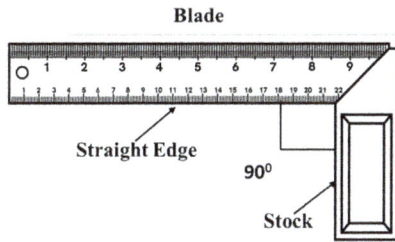

(b) Try Square

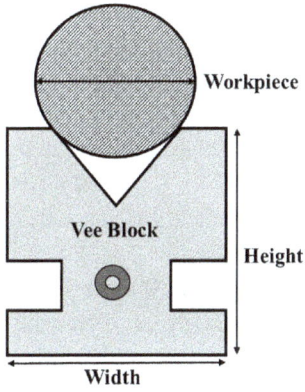

(c) Vee Block

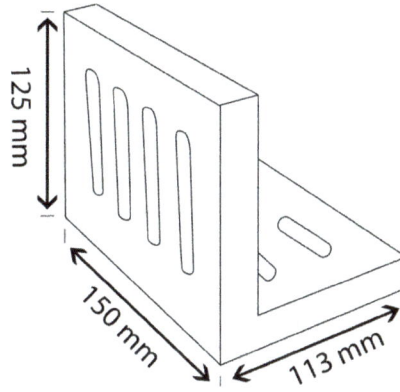

(d) Angle Plate

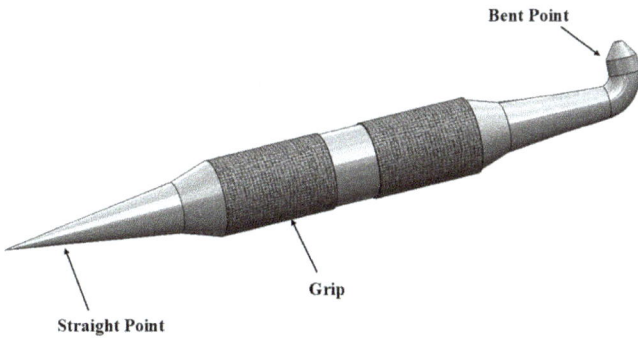

(e) Scriber

Figure 2.31: (a) Surface plate; (b) try square; (c) vee block; (d) angle plate; (e) scriber.

d) **Angle plate**: It has two plan surfaces at right angle to each other. This is made of gray cast iron. This is used for supporting work in a perpendicular position with the help of surface plate. There are many slots on it which provide the facility to hold the work by bolt and damp.

e) **Scriber**: It is used like a pencil to scratch or scribe line on metal. This is made of hardened steel with length about 150–300 mm and thickness 3–5 mm.

viii. **Punch:** Punch is used to locate mark on work. There are two types of punches in common, as shown in Figure 2.32:

a) **Prick punch:** It has sharp pointed tool and tapered point of punch is angled at 40°. It is used to make small marks on layout lines to get long lasting mark. Basic view of Prick punch has been sown in figure 2.32 (e).

b) **Center punch:** A center punch has a pointed tip with an angle of 60°, compared to the 40° angle of a prick punch. The indentation made by a center punch is typically used to guide drills for accurate hole making, as illustrated in Figure 2.32(a).

c) **Pin punch**: Used to drive pins out of holes, it has a flat tip that fits into the pin's hole. Sketch of pin punch has been shown in figure 2.32 (b).

d) **Starting punch**: "Starting punch" is a type of punch tool used to start the removal of pins and bolts. See figure 2.32 (d).

e) **Hollow shank gasket punch**: A hollow shank gasket punch is a specialized tool used for cutting holes in gasket materials.

f) **A lining punch:** A line punch, also known as a layout punch, is a specialized tool used in metalworking and woodworking to mark straight lines on a surface. It helps in transferring measurements and marking lines for cutting, drilling, or other fabrication processes. The line punch is designed to create a visible, precise line that can guide subsequent operations.

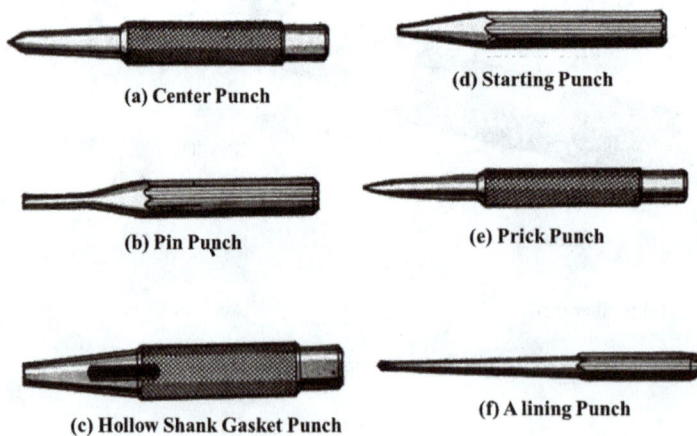

(a) Center Punch

(d) Starting Punch

(b) Pin Punch

(e) Prick Punch

(c) Hollow Shank Gasket Punch

(f) A lining Punch

Figure 2.32: Types of punch.

2.4.2 Measuring instruments

Some commonly measuring instruments, shown in Figure 2.33(a) to (d), used in bench work and fitting are given below:

a) **Steel rule:** It is used to measure linear measurement of work. It is made of hardened steel. It consists of s strip with line graduation etched on it. At interval of fraction of a standard unit, it has an accuracy of range 1.0–5 mm. On the basis of graduation it can be manufactured in various sizes and styles.

b) **Divider:** It is used to scribe circle and general layout work.

c) **Micrometer**
   - Used to measure the diameters of shafts, screw threads, holes, and thickness of parts.
   - Provides an accuracy of 0.01 mm.
   - The least count of an instrument refers to the smallest measurable value.
   - For a micrometer, the least count is typically 0.01 mm.

   Least count = Number of divisions on the thimble/pitch of the screw

d) **Vernier caliper**
   - Measures both internal and external diameters of shafts and the thickness of components.
   - Provides an accuracy of approximately 0.02 mm.
   - Works using a Vernier scale, which utilizes the difference between two nearly identical scales for precise measurement.
   - Components:
     - Main scale with fixed graduations.
     - Two measuring jaws for internal and external measurements.
     - Vernier head with an engraved Vernier scale.
     - Auxiliary head with a Vernier clamp, adjusted using a micrometer screw.
     - Knurled screw to lock the Vernier head and auxiliary head to the main scale.

   Least count = Smallest division on main scale/total number of divisions
                  on Vernier scale

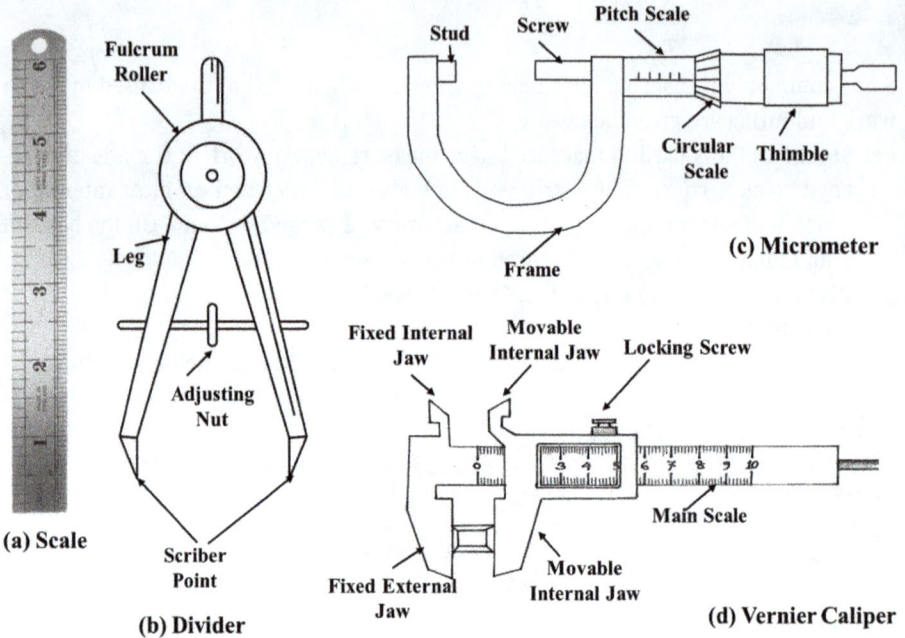

Figure 2.33: Various measuring instrument.

Questions

Long answer-type questions

1. Explain the rolling process and its classification. Discuss the differences between hot rolling and cold rolling, including the advantages and applications of each.

2. Describe the rolling mill arrangements and the role of different types of rolls (e.g., two-high, three-high, four-high, and cluster mills) in the metal rolling process. How do these arrangements impact the quality and efficiency of rolling operations?

3. What are the key defects encountered in the rolling process? Explain their causes, preventive measures, and the impact of these defects on the final product.

4. Differentiate between open-die and closed-die forging. Describe the processes, advantages, limitations, and typical applications of each type.

5. Explain the significance of the forging temperature and the effects of improper temperature control on the quality of forged products.

6. Discuss the various methods used for forging defect detection and correction. How can process parameters be optimized to minimize defects?

7. Describe the extrusion process in detail including direct and indirect extrusion. Compare the two methods in terms of energy efficiency, product quality, and applicability.

8. Discuss the role of die design in the extrusion process. How do factors like die shape, material, and lubrication affect the extrusion process and the final product quality?

9. Explain the difference between hot extrusion and cold extrusion. What are the advantages and challenges associated with each method? Provide examples of products made using these processes.

10. What are the various operations involved in sheet metal working? Discuss the significance of operations like shearing, blanking, and punching, and explain how these processes are integrated into manufacturing.
11. Discuss the factors affecting the selection of sheet metal for a specific application. How do material properties, thickness, and process limitations influence the choice of sheet metal?
12. Explain the concept of bending in sheet metal working. Discuss the factors that influence the bending process such as bend radius, material properties, and springback effect.
13. Describe the deep drawing process and its applications. What are the critical parameters that influence the success of deep drawing operations, and how can they be controlled to prevent defects like wrinkling and tearing?
14. Explain the importance of blank holder force in the deep drawing process. How does it affect the quality of the final product, and what methods can be used to optimize this force?
15. Discuss the limitations and challenges of the deep drawing process. How can modern techniques, such as hydroforming, be used to overcome these challenges?
16. Explain the wire drawing process including the steps involved and the types of equipment used. How does the process differ when drawing different materials (e.g., steel vs. copper)?
17. What are the common defects in the wire drawing process and what are their causes? Discuss the methods used to detect and correct these defects.
18. Describe the importance of bench work and fitting in the manufacturing process. How do precision and accuracy in fitting operations impact the overall quality of the final product?
19. Explain the various tools and techniques used in bench work and fitting. Discuss the role of each tool in the process and the skills required to use them effectively.
20. Discuss the safety considerations in bench work and fitting operations. What measures should be taken to ensure a safe working environment, and how can workers be trained to follow these safety protocols?

Short answer-type questions
1. What is the primary purpose of the rolling process in metal forming?
2. Differentiate between hot rolling and cold rolling.
3. Explain the significance of the roll gap in the rolling process.
4. What is the role of friction in the rolling process?
5. Name two products that are commonly manufactured using the rolling process.
6. What is the main difference between open-die forging and closed-die forging?
7. Explain the term "forging force" and its importance in the forging process.
8. What is the significance of grain flow in forged components?
9. Name two types of forging hammers used in the forging process.
10. Why is lubrication important in forging operations?
11. Differentiate between direct extrusion and indirect extrusion.
12. What are the advantages of using the extrusion process in manufacturing?
13. Explain the term "extrusion ratio" and its impact on the process.
14. What materials are commonly used in the extrusion process?
15. What is the purpose of the die in the extrusion process?
16. What is the significance of the punch and die in sheet metal working?
17. Define the term "blanking" in sheet metal operations.
18. What is the purpose of using a press brake in sheet metal working?
19. Explain the difference between shearing and bending in sheet metal working.
20. What is the role of springback in sheet metal forming operations

MCQ

1. Which of the following is a key advantage of the rolling process in metal forming?   **[GATE 2020]**
    (A) Produces complex shapes easily
    (B) Increases the hardness of the material
    (C) Produces uniform thickness with good surface finish
    (D) Is suitable for non-metallic materials

2. In forging, the main advantage of using hot forging over cold forging is: **[IES 2018]**
    (A) Better surface finish
    (B) Reduced oxidation
    (C) Easier deformation of metal
    (D) Increased dimensional accuracy

3. Extrusion is a process used to produce: **[SSC JE 2019 ME]**
    (A) Hollow parts
    (B) Thin sheets
    (C) Long lengths of uniform cross-section
    (D) High surface finish

4. In deep drawing, the ratio of the blank diameter to the punch diameter is known as: **[GATE 2017]**
    (A) Drawing ratio   (B) Reduction ratio   (C) Anisotropy ratio   (D) Forming limit

5. Which metal forming process is primarily used to produce thin wires from thicker rods?   **[RRB JE 2018, ME]**
    (A) Rollin   (B) Forging   (C) Extrusion   (D) Wire drawing

6. Which of the following operations is not typically associated with sheet metal working?   **[IES 2019]**
    (A) Bending   (B) Drawing   (C) Forging   (D) Blanking

7. During the wire drawing process, which factor primarily affects the reduction in wire diameter?
                                                                                        [GATE 2021]
    (A) Die material   (B) Lubrication   (C) Die angle   (D) Wire material

8. In bench work and fitting, which tool is primarily used for measuring small gaps or checking the flatness of a surface?   **[SSC JE 2020 ME]**
    (A) Vernier caliper   (B) Micrometer   (C) Feeler gauge   (D) Surface plate

Answers
1(C) 2(C) 3(C) 4(A) 5(D) 6(C) 7(C) 8(C)

Chapter 3
Machining processes and their applications

3.1 Introduction to cutting tools

Tools are very important for proper cutting or machining. Dimensional accuracy and fit can be achieved by proper tool geometry. With the proper tool geometry, machining or cutting of metal becomes efficient and economical. Cutting tools are classified as:

i.   Single-point cutting tool
ii.   Multi-point cutting tool
iii. Abrasive cutting

A single-point cutting tool is a tool that has a single cutting edge. During a machining operation, only one cutting edge participates in metal removal or cutting. On the other hand, a multi-point cutting tool is a tool that has multiple cutting edges. During cutting, multiple cutting edges participate in metal removal. The metal removal rate is faster when a multi-point cutting tool is used.

In abrasive cutting, abrasive particles are used to remove metal. In this process, abrasive particles strike the surface of the metal with high velocity, and due to the higher velocity of the abrasive particles, the metal is removed.

3.1.1 Nomenclature of single-point cutting tools

Cutting tools are made in specific geometrical shapes at the cutting point. There are some predefined shapes of the cutting edge that have been developed, and various angles are formed at the cutting point. These angles are defined by the following system of tool nomenclature. These systems are:

a) American standards association system
b) Continental (German) system
c) British maximum rake system

3.2 Basic principles of lathe machines

In a lathe machine, the cutting motion is achieved by rotating the workpiece, while the feed motion is provided by the linear movement of the cutting tool.

–   Longitudinal feed: The tool moves axially along the workpiece.
–   Cross feed: The tool moves radially toward or away from the workpiece.

https://doi.org/10.1515/9783112205891-003

a) **Power transmission in a lathe machine**
- The workpiece rotates and receives power from an electric motor.
- Power is transmitted through a belt-and-pulley system, a clutch, and a speed gearbox.
- The speed gearbox divides the input speed into multiple speed options using a cluster gear system.

b) **Tool feed mechanism**
The feed mechanism is responsible for transmitting power to the carriage, ensuring precise movement of the cutting tool during machining. Initially, power is transmitted by a gear train to the quick-change gear box, which, in turn, regulates the tool movement per revolution of the spindle. It also ensures smooth and controlled movement of the tool, improving machining accuracy and efficiency.

- **Working of the feed mechanism**
  - Power transmission:
    - Power is initially transferred through a gear train connected to the lathe spindle.
    - This gear train drives the quick-change gearbox.
  - Quick-change gearbox:
    - Regulates the feed rate, determining the movement of the cutting tool per revolution of the spindle.
    - Allows quick and easy selection of different feed rates for various cutting operations.
  - **Feed rod and leadscrew**
    - The feed rod transfers motion from the gearbox to the carriage for automatic feed.
    - The leadscrew is used for thread cutting, ensuring precise tool movement along the workpiece.
- The automatic feed motion of the tool is controlled by the feed gearbox and the apron gear system.
- **The feed rod transmits motion either to:**
  i. A pinion-and-rack system, which moves the tool longitudinally along the workpiece.
  ii. A cross-slide screw moves the tool radially for cross-feed.

This mechanism allows precise cutting, shaping, and machining of the workpiece in lathe operations.

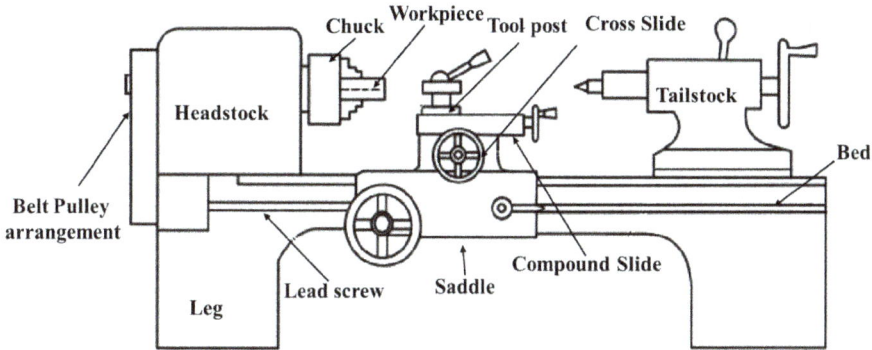

Figure 3.1: Parts of a lathe machine.

3.2.1 Parts of a lathe machine

The main parts of the lathe machine are illustrated in Figure 3.1 and described below.

i. **Head stock:** The headstock is a critical part of a lathe machine that houses the main spindle, speed gearbox, and power transmission components. It is located on the left side of the lathe bed and is responsible for driving the workpiece rotation. It plays a vital role in power transmission, speed control, and workpiece holding, making it one of the most important parts of a lathe machine.
– **Functions of the headstock:**
   a) Holds and rotates the workpiece:
      – The spindle inside the headstock securely holds the chuck, faceplate, or collet, which grips the workpiece.
      – The spindle is driven by an electric motor through a belt-and-gear system.
   b) Speed control:
      – The speed gearbox inside the headstock provides variable spindle speeds.
      – Speed changes are achieved by shifting gears, pulleys, or belt positions.
   c) Power transmission:
      – The power from the motor is transmitted through a belt-and-pulley system or gear train to the spindle.
      – Some lathes have a clutch system to engage or disengage power transmission.
   d) Supports other attachments:
      – Certain lathes allow the mounting of attachments, such as a taper-turning attachment or thread-cutting gears, on the headstock.

– **Key components of the headstock:**
 a) Main spindle: Rotates the workpiece and holds the chuck or faceplate. The spindle of the headstock is hollow throughout its length and is supported on two bearings. The outer surface of the spindle is generally threaded to assist in mounting a chuck or a face plate. The headstock also contains some sub-parts, such as the high-speed and low-speed lever, and the tumbler lever.
 b) Gearbox: Controls the spindle speed using different gears.
 c) Bearings: Support and ensure the smooth rotation of the spindle.
 d) Belt and pulley system: Transfers power from the motor to the spindle.
 e) Clutch and brake mechanism (in advanced lathes): Allows smooth engagement and stopping of the spindle rotation.

ii. **Tailstock:** The tailstock is a movable component located on the right side of a lathe machine. It provides support for long workpieces during machining. It is a casting that can move along the bed way to accommodate different lengths of workpieces. The upper part of the tailstock accommodates a hollow barrel; one end of this barrel holds the dead center. This part is also used to support the feed tool to perform various operations such as reaming, drilling, etc. The tailstock is illustrated in Figure 3.2(a).

– **Main components of the tailstock:**
 a) Base: Mounted on the lathe bed and can slide along it.
 b) Clamp or locking mechanism: Secures the tailstock in place.
 c) Spindle (quill): A hollow shaft that moves forward and backward to hold the tool.
 d) Handwheel: Used to move the spindle in and out.
 e) Offset adjustment: Allows minor lateral movement for taper turning.

iii. **Bed:** The bed is the base structure of a lathe machine, providing the foundation for the entire machine and holding components such as the headstock, tailstock, and carriage. The bed of the lathe machine is made of cast iron. It has V and U-type ways to facilitate the movement of the carriage and tailstock.

iv. **Lead screw:** The lead screw is a long-threaded rod running along the front of the lathe bed, essential for thread cutting and automatic feed. Functions of the lead screw are to transfer rotary motion from the spindle to the carriage, enabling automatic longitudinal feed, and to ensure uniform feed and accurate cutting operations. It ensures precise movement of the carriage along the lathe bed in synchronization with the spindle rotation. A train of gears is used to connect the lathe spindle to the leadscrew and the lead screw to the lathe carriage. This carriage, along with its cutting tool, can be forced to move a set distance for every revolution of the lathe spindle.

v. **The carriage**: The carriage depicted in Figure 3.2(b) is an essential component of the lathe, responsible for holding, guiding, and feeding the cutting tool against the workpiece. It moves along the bed between the headstock and tailstock. The carriage

assembly plays a crucial role in precision machining, providing controlled movement of the cutting tool along multiple axes for various lathe operations.
– **The carriage consists of several sub-components:**
  a) Apron:
     – Located at the front of the carriage and attached to the saddle.
     – Contains split nuts, gears, and clutches to transmit motion from the feed rod to the carriage during thread cutting.
  b) Saddle:
     – Mounted directly on the lathe bed ways, allowing longitudinal movement.
     – Houses the cross-slide, compound rest, tool post, and tool holder.
  c) Cross slide:
     – Positioned on top of the saddle to provide transverse movement for the cutting tool.
     – Can be operated manually using a handwheel or automatically via a power-feed mechanism.
  d) Compound rest:
     – Mounted on the cross slide with a circular base graduated in degrees.
     – Can be swiveled to various angular positions, essential for taper turning and precision cuts.
  e) Tool post:
     – The topmost part of the carriage, mounted on the compound rest.
     – Designed to securely hold various cutting tools in the tool holder.
     – Tool holder: A component that firmly grips the cutting tool, ensuring stability during machining.

vi. **Legs:** Lathe legs support the entire machine's weight and provide stability. They are firmly secured to the floor using foundation bolts to absorb vibrations. Made of cast iron, they ensure durability and rigidity.

3.2.2 Operations performed on a lathe machine

Figure 3.3 illustrates the numerous operations performed on a lathe machine which are discussed below:
i. **Turning:** This is the process of removing excess material from the outer surface of a workpiece to achieve a smooth, finished surface. Turning operations are generally classified into two main types.
ii. **Facing:** This operation involves removing material from the end of a cylindrical workpiece, effectively reducing its length.

iii. **Plain turning:** In this process, the cross-sectional area of a cylindrical workpiece is reduced. The cutting tool is set to a specific depth using the cross-slide and is then moved parallel to the spindle axis to remove excess material, resulting in a smooth cylindrical surface.

iv. **Taper turning:** This operation creates a conical shape by gradually decreasing the diameter of a cylindrical workpiece.

v. **Step turning:** This process involves machining a workpiece to produce multiple steps with varying diameters.

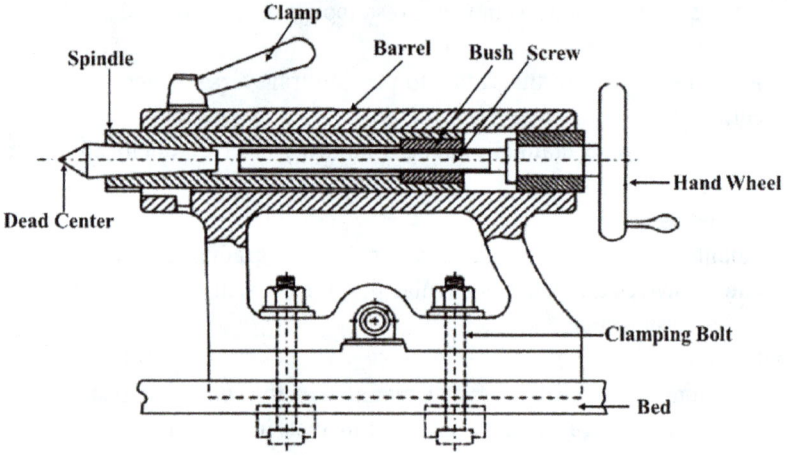

(a)

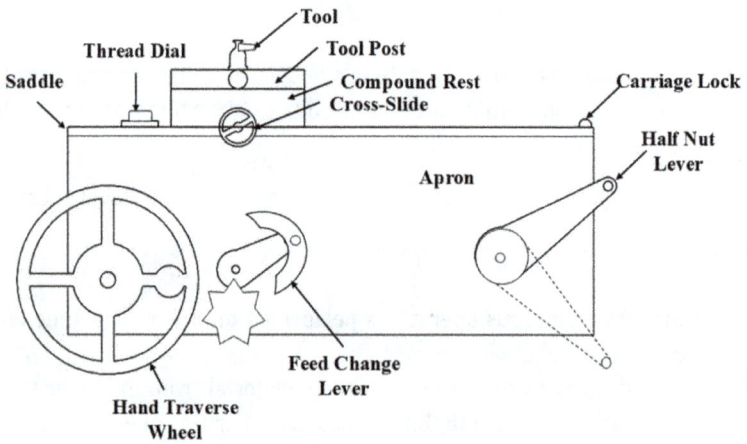

(b)

Figure 3.2: (a) Tailstock, (b) carriage assembly.

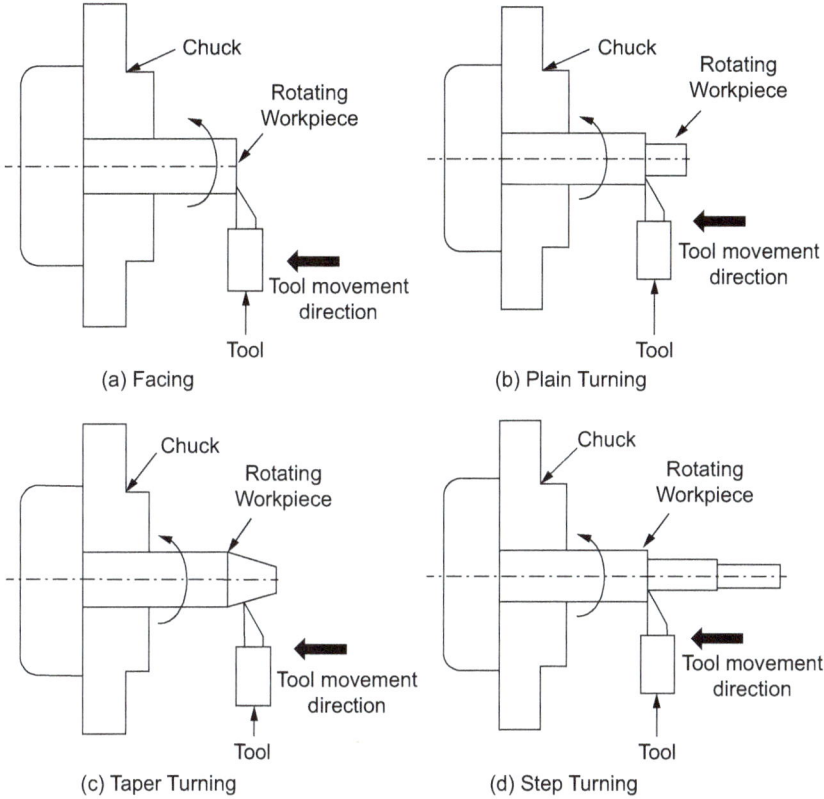

Figure 3.3: (a) Facing, (b) plain turning, (c) taper turning, (d) step turning.

In addition, this operation and several other operations can be performed on a lathe, such as drilling, boring, reaming, milling, knurling, grooving, necking, parting off, and grinding.

3.3 Basic description of another machine tool

3.3.1 Shaper

A shaper, as shown in Figure 3.5, is a machine equipped with a reciprocating cutting tool that moves in a straight line to perform machining operations. A flat surface is produced as the workpiece moves across the path of this reciprocating tool, as shown in Figure 3.4. The machine operates using power supplied by an electric motor.

The shaper consists of a base and a frame that support a horizontal ram. The ram carries the cutting tool and moves back and forth, covering the required stroke length.

A clapper box tool holder is mounted on the ram through a pivot at its upper end, allowing controlled cutting action.

A shaper machine is primarily used for machining flat surfaces, grooves, and other contours. The shaping process involves the tool making a forward stroke to remove material, while the return stroke remains idle.

– **Working principle of shaper**

   a) The shaper machine operates on the **quick-return mechanism**, which converts the rotary motion of the crank into the reciprocating motion of the cutting tool.

   b) It is primarily used to machine horizontal and vertical flat surfaces using a single-point reciprocating tool.

   c) The reciprocating motion of the tool is achieved by transforming the rotary motion of the drive system into linear movement.

   d) The quick return mechanism is designed so that the return stroke is faster than the cutting stroke, improving efficiency and reducing idle time.

   e) The quick return mechanism consists of a rotating crank, driven by a motor at a uniform speed, and connected to a reciprocating arm through a sliding block.

   f) The sliding block allows the adjustment of the stroke length via a screw mechanism.

   g) The stroke length can be varied by changing the length of the crank.

   h) The ratio of the return stroke to the cutting stroke is approximately 2:3, enhancing operational efficiency.

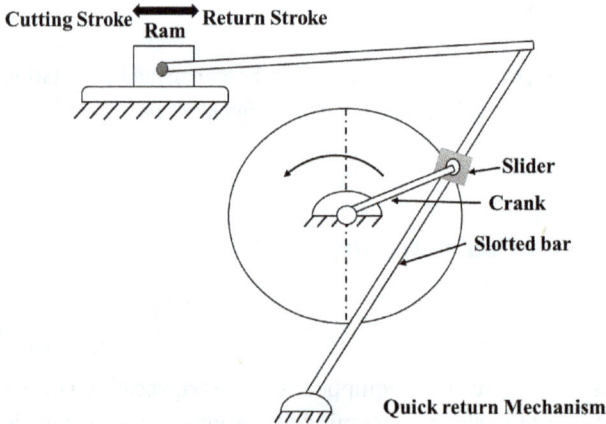

Figure 3.4: Working principle of shaper machine.

– **Parts of a shaper**
Here are the main parts of a shaper machine, as shown in Figure 3.5:
  i.   **Base:** Supports the entire machine and absorbs vibrations.
  ii.   **Column**: Houses the drive mechanism and supports the reciprocating parts.
  iii. **Ram**: Holds the cutting tool and moves back and forth in a straight line.
  iv.   **Table**: Supports the workpiece and provides movement for machining.
  v.   **Cross rail**: Supports the table and allows vertical adjustment.
  vi.   **Saddle**: Mounted on the cross rail, providing lateral movement of the table.
  vii. **Tool head**: Holds and controls the cutting tool, allowing angular adjustments.
  viii. **Clapper box**: Holds the tool post and allows free movement of the tool during the return stroke.
  ix.   **Feed mechanism**: Controls the movement of the table for the cutting operation.
  x.   **Quick return mechanism**: Converts rotary motion into reciprocating motion and ensures a faster return stroke.
  xi.   **Driving motor**: Provides power to drive the shaper mechanism.
  xii. **Handwheel**: Used for manual adjustments of table position and feed.

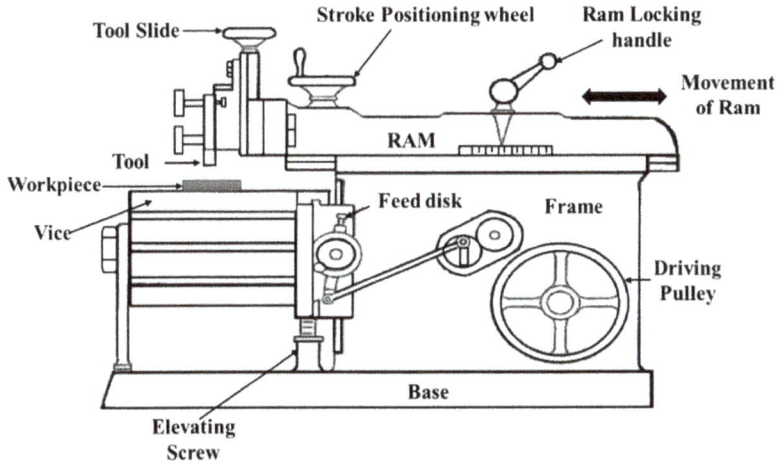

Figure 3.5: Basic components of a shaper machine.

– **Shaper operation**
The various operations carried out on the shaper machine are shown in Figure 3.6 and discussed below.
  a) Horizontal cutting: Used to machine flat surfaces parallel to the table. The workpiece is fed horizontally while the tool reciprocates in a straight line.
  b) Vertical cutting: This involves machining flat surfaces perpendicular to the table. It is achieved by adjusting the tool head to a 90-degree angle.

c) Angular cutting: This process is performed by tilting the tool head or workpiece to cut inclined surfaces at a desired angle. It is commonly used for dovetail and chamfer cutting.

d) Irregular cutting: Used to produce contoured or curved surfaces. This is achieved using a form tool or by combining vertical and horizontal feed movements to generate the required shape.

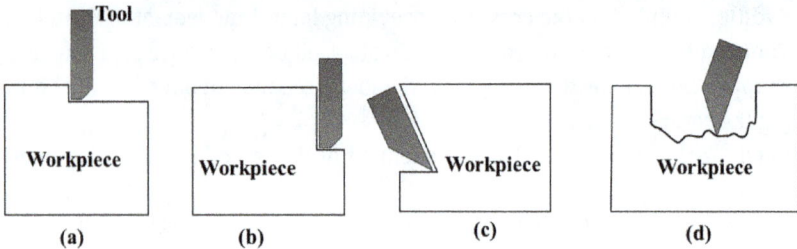

Figure 3.6: (a) Horizontal cutting, (b) vertical cutting, (c) angular cutting, (d) irregular cutting.

- **Types of surfaces commonly machined by shaping and planning:** Figure 3.7 depicts various machined surfaces, which are addressed in more detail below.

a) Flat horizontal surfaces: The most common type of surface produced using shaping and planning, achieved by reciprocating the tool in a straight path.

b) Flat vertical surfaces: Machined by adjusting the tool head vertically to create perpendicular cuts.

c) Inclined or angular surfaces: Produced by tilting the tool head or workpiece to the required angle, and they are commonly used for dovetail joints and chamfers.

d) Curved or contoured surfaces: Created using form tools or a combination of feed motions, these are often required in decorative and specialized machining applications.

e) Slots and grooves: Shaping and planning machines are used to cut keyways, T-slots, and dovetail slots on workpieces.

f) Steps and shoulders: These are machined by controlling the depth of cut at different levels, which is often required in component assembly.

g) V-grooves and notches: Produced by using specially designed cutting tools for applications such as guideways and mating parts.

h) Interlocking or matching surfaces: Used in precision engineering where parts must fit together perfectly, such as slides and machine beds.

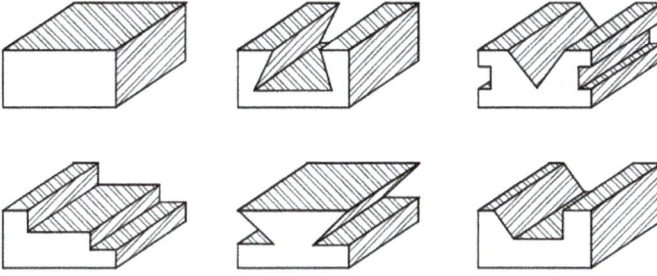

Figure 3.7: Types of surfaces commonly machined by shaping and planning.

3.3.2 Planer

Planners are similar to shapers, where the work reciprocates instead of the tool. The feed motion is given to the tool mounted on a cross-rail head. A planer is shown in Figure 3.8. It consists of a large bed that carries guide ways for the work table, which holds the workpiece to reciprocate. The table movement is actuated by a variable-speed drive using a rack-and-pinion arrangement or a hydraulic system. The housing carries three heads: two on the vertical side columns and the third on the cross-rail. The tool is given intermittent feed motion. The cross-rail can be raised or lowered to accommodate jobs of various sizes.

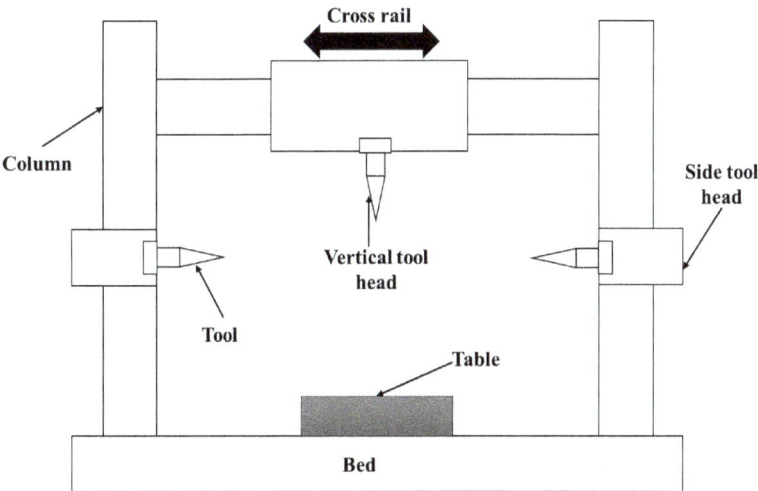

Figure 3.8: Planer machine.

The key differences between a shaper and a planer are outlined in Table 3.1, which compares the two machines based on working principles, size, applications, and typical usage in machining operations.

Table 3.1: Comparison between shaper and planer.

Shaper	Planer
Lighter machine planer	Heavy and robust machine
Tools reciprocate against the fixed workpiece	Work piece reciprocate against the fixed tools
Limited to machines for small workpieces	Used for large and heavy workpieces
Large depths of cut and coarse feeds are not possible	Large depths of cut and coarse feeds are possible
Usually accommodate only one tool	It can accommodate multiple tools
Feed is given to the work during the idle stroke of the ram	Feed is given to the tool during the idle stroke

3.3.3 Milling machine

Milling machines are utilized to manufacture components with both flat and curved surfaces. They are capable of producing intricate shapes that cannot be easily achieved with other machine tools. During the milling process, the workpiece is typically fed into a rotating cutting tool called a milling cutter. The cutter has evenly spaced teeth along its periphery, which engage with the workpiece to remove material.

Milling machines are used for machining flat surfaces, contoured surfaces, surfaces of revolution, and internal and external threads. Complicated shapes that are not easy to produce on other machines can be produced on a milling machine. Due to their high accuracy and production rates, milling machines have replaced the shaper and slotter. A milling machine uses a multi-point cutting tool called a milling cutter. A milling cutter is a multi-point cutting tool that has multiple cutting edges on its circumference. The cutting edges (teeth) are equally spaced on the periphery of the tool. In milling operations, the workpiece is fed into the rotating milling cutter. These cutting edges come into contact with the workpiece and remove the metal. In some special milling machines, the workpiece remains stationary while the cutter is fed into the workpiece. Due to the multiple cutting edges, the metal removal rate is faster than in other machining processes. Various types of cutters are used in milling machines, such as end mill cutters, side cutters, and face cutters, etc.

– **Types of milling machines**
Depending on the axis of rotation of the milling cutter, milling machines are of two major types. Another feature is their knee-and-column construction. These are general-purpose machines. Thus, we have two classes of machines. The parts are shown in Figure 3.9.

– **Main parts of a column and knee-type milling machine**
i.   **Base**: Supports the machine and absorbs vibrations.
ii.   **Column**: Houses the spindle, motor, and feed mechanism.
iii. **Knee**: Moves vertically along the column to adjust the table's height.
iv.   **Saddle**: Mounted on the knee, allowing horizontal movement of the table.
v.   **Table**: Holds the workpiece and moves in different directions for machining.
vi.   **Overhanging arm**: Supports the arbor for horizontal milling.
vii. **Arbor**: Holds and drives the milling cutter in horizontal milling machines.
viii. **Feed mechanism**: Controls the movement of the workpiece toward the cutter.

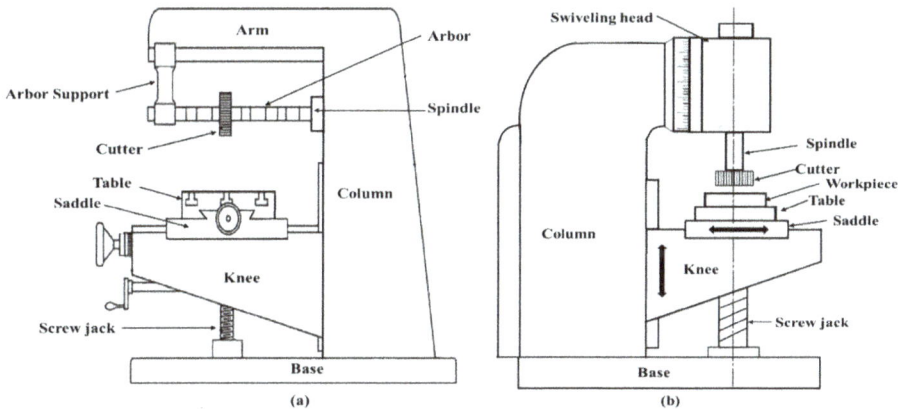

Figure 3.9: (a) Horizontal knee-and-column type milling machine. (b) Vertical knee-and-column type milling machine.

– **Milling processes can be broadly classified as** shown in Figure 3.10.
i.   **Plain milling**: Cutting flat surfaces parallel to the milling cutter's axis. See figure 3.10(a).
ii.   **Face milling**: Machining flat surfaces using a cutter mounted on a spindle. See Figure 3.10(b).
iii. **Side milling**: Cutting vertical surfaces using a side-mounted cutter.
iv.   **End milling**: Machining pockets, slots, and intricate contours with an end mill cutter. see figure 3.10(f).
v.   **Angular milling**: Cutting angled surfaces by tilting the cutter or the workpiece. see figure 3.10(d).

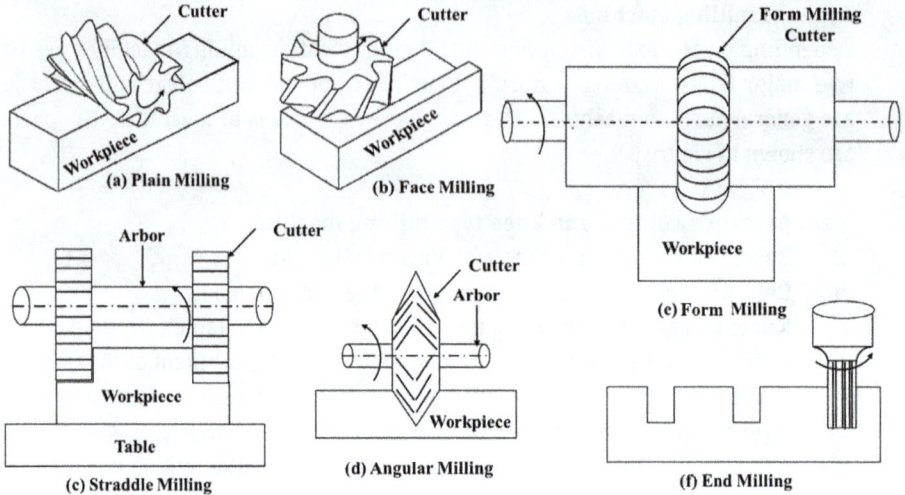

Figure 3.10: Various types of milling processes.

   vi.   **Form milling**: Producing irregular contours using specially shaped cutters. see figure 3.10(e).

  vii.   **Straddle milling**: Cutting two parallel surfaces simultaneously with two side cutters. See figure 3.10(c).

 viii.   **Slotting**: Creating narrow slots or keyways in a workpiece.

   ix.   **Gear milling**: Machining gear teeth using form cutters or gear-hobbing techniques.

–   **Peripheral or slab milling can be further subdivided into**
  1.   Up milling
  2.   Down milling

1.   **Up milling:** When the direction of rotation of the milling cutter and the direction of feeding of the workpiece are in opposite directions, it is called up milling, as shown in Figure 3.11(a). It is also called conventional milling. In this type of milling, rigid clamping is required due to the tendency of lifting the work. In this process, the work is fed against the motion of the cutter. Chip formation starts with minimum thickness and increases to maximum thickness at the end.

2.   **Down milling:** When the direction of rotation of the milling cutter and the direction of feed of the workpiece are the same, it is called down milling, as shown in Figure 3.11(b). This is also known as climb milling. In this type of milling operation, chip formation starts with maximum thickness and decreases to minimum thickness at the end. In this work, piece is fed in the direction of the cutter. Rigid clamping is not required in this case.

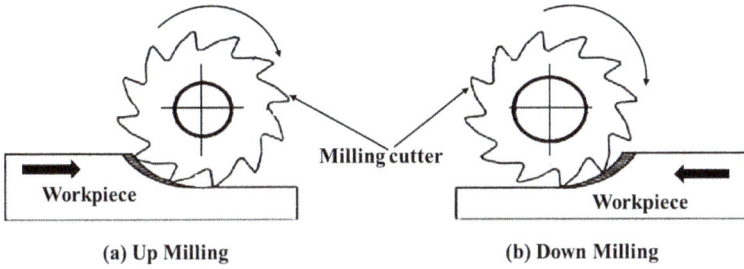

Milling cutter

(a) Up Milling
(b) Down Milling

Figure 3.11: (a) Up milling. (b) Down milling.

The differences between up milling and down milling are clearly illustrated in Table 3.2, which compares the two methods based on tool rotation, chip formation, surface finish, and their suitability for different machining applications.

Table 3.2: Differences between up milling and down milling.

Up milling	Down milling
Workpiece is fed against the rotation of the cutter	Workpiece is fed in the same direction as that of the cutter's rotation
Chip thickness starts from zero and ends at maximum	Chip thickness starts from maximum and ends at minimum
Cutting force acting in an upward direction	Cutting force is acting in a downward direction
Chip removal is not easy	Easy chip removal
Thick jobs are machined	Thin jobs are machined
Strong clamping is needed to counter act upward cutting	Less clamping force is needed as cutting helps in pressing the job
Difficult to provide cooling	Easy to provide cooling
Cutting forces start from zero at the beginning and increase to a maximum at the end of cutting by the tooth	Cutting force is at its maximum at the beginning and reduces to its minimum at the end of cutting by the tooth

3.3.4 Grinding

a) Definition: Grinding is a metal-cutting process in which material is removed from the workpiece using a high-speed rotating abrasive tool called a grinding wheel, as shown in Figure 3.12.
b) Grinding wheel speed: The grinding wheel operates at a velocity ranging from 10 to 80 m/s.
c) Workpiece feed rate: The workpiece is fed past the grinding wheel at a speed of 0.2 to 0.6 m/s.
d) Similarity to up-milling: Grinding bears some resemblance to the up-milling process in terms of material removal.
e) Abrasive cutting points: Unlike milling cutters, the cutting points on a grinding wheel are irregularly shaped and randomly distributed, making the process different from conventional machining.

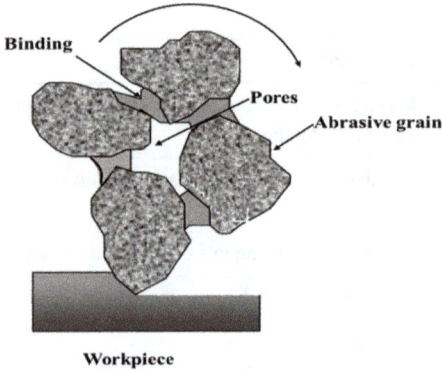

Figure 3.12: Elements of a grinding wheel.

- **Abrasive and bonds:** Abrasives are fine particles with sharp cutting points. These abrasive particles are bonded together by organic binders. During the cutting action, the sharp points of abrasive particles get damaged, so they cannot cut the workpiece again. Consequently, cutting forces cause the bond to fracture and fail in order to expose a new sharp abrasive particle, which participates in cutting while the damaged particle is removed. The removal of damaged abrasive particles from the wheel is called friability. Therefore, the bonding should not be very strong; otherwise, the damaged particles will remain bonded and will not machine the surface properly. The strength of the bonding is normally termed the grade of the wheel.

In grinding, the cutting action is similar to the chip removal process in milling. The chips formed during grinding are very fine, approximately one-tenth the thickness of a human hair.

- **Classification of grinding machines**

i. **Types of grinding machines based on accuracy:** Grinding machines can be broadly classified into two categories based on their accuracy and purpose:

1. **Rough grinders**
   - Used for rapid material removal rather than for precision.
   - Suitable for heavy stock removal, deburring, and cleaning operations.
   - Produces coarse surface finishes with lower accuracy.
   - Commonly used in foundries, welding shops, and fabrication industries.

- **Types of rough grinders**
   a) Bench grinder: A stationary grinder used for sharpening tools and for rough grinding of small parts.
   b) Pedestal grinder: Similar to a bench grinder, but mounted on a stand for larger workpieces.
   c) Swing frame grinder: Used for heavy-duty grinding in casting and forging industries.
   d) Belt grinder: Uses an abrasive belt for grinding and finishing operations.
   e) Flexible shaft grinder: Equipped with a flexible shaft for grinding hard-to-reach areas.

2. **Precision grinders**
   - Used for achieving high accuracy and fine surface finishes.
   - Removes very small amounts of material to achieve precise dimensions.
   - Commonly used in tool-making, aerospace, and automotive industries.

- **Types of precision grinders**
   a) Surface grinding machine: Used for grinding flat surfaces with high precision.
   b) Cylindrical grinding machine: Used for grinding external and internal cylindrical surfaces.
   c) Centerless grinding machine: Used for precision grinding of cylindrical parts without a center.
   d) Tool and cutter grinder: Used for sharpening cutting tools such as drills and milling cutters.
   e) Thread grinding machine: Used for grinding precise threads on screws and bolts.
   f) Jig grinder: Used for grinding precise holes and complex contours.
   g) Belt grinding machines: Use an abrasive belt instead of a grinding wheel.
   h) Special-purpose grinding machines
      - Gear grinding machine: Used for grinding gear teeth with high precision.
      - Creep-feed grinding machine: Used for deep and slow material removal in tough materials.

a) **Surface grinding: overview, process, and applications**
  - Definition: Surface grinding, as shown in Figure 3.13, is a precision machining process used to produce smooth, flat surfaces on metallic and non-metallic workpieces by removing material using an abrasive grinding wheel.

  - **Working principle**
    - The workpiece is clamped on a table (magnetic chuck for ferrous materials or mechanical clamps for non-ferrous materials).
    - A rotating abrasive grinding wheel removes material from the surface of the workpiece.
    - The table moves longitudinally and/or transversely, allowing uniform material removal.
    - The grinding wheel moves down gradually, making light cuts to achieve a high level of accuracy and surface finish.

  - **Types of surface grinding machines**
    a) Horizontal spindle reciprocating table: The workpiece moves back and forth beneath a rotating grinding wheel.
    b) Horizontal spindle rotary table: The workpiece rotates while the grinding wheel moves radially.
    c) Vertical spindle reciprocating table: The grinding wheel is positioned vertically and is suitable for larger surfaces.
    d) Vertical spindle rotary table: Used for precision surface grinding of round or irregularly shaped workpieces.

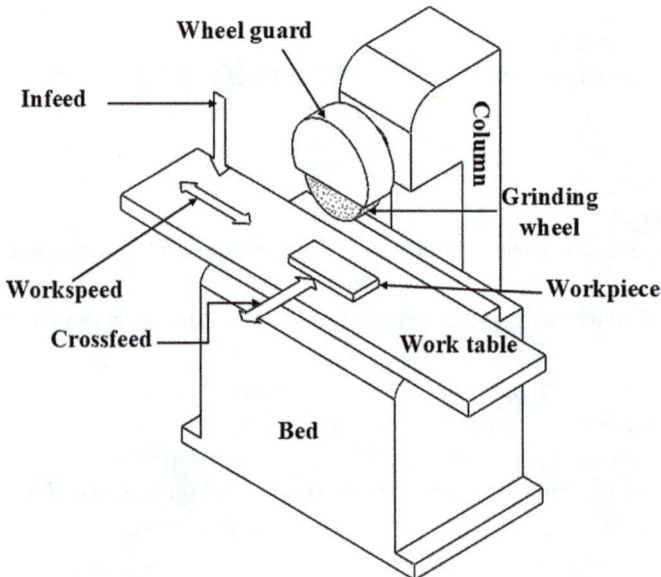

Figure 3.13: Detail of surface grinding.

- **Grinding operation is performed to**
  i. Machine hard surfaces: Grinding is used for machining hardened materials that cannot be easily cut using conventional machining methods.
  ii. Impart better surface finish: Produces smooth, high-quality surface finishes, enhancing the appearance and functionality of the workpiece.
  iii. Sharpen cutting tools: Used for reconditioning and sharpening cutting tools, such as drills, milling cutters, and lathe tools, to maintain precision.
  iv. Finish the workpiece within close tolerance: Achieves high accuracy by removing small amounts of material to meet precise dimensional requirements.

- **Grinding wheel specification:** Grinding wheels are specified on the basis of their elements, as shown in Figure 3.14.
  i. Abrasive type: There are three types of abrasives commonly used in commercial grinding wheels, viz. aluminum oxide, silicon carbide, and diamond. The figure shows details regarding the grinding wheel.
  ii. Grain size: The grain size of the abrasive particles used can be grouped into coarse grains (10–24), medium-sized grains (30–60), fine grains (80–180), and very fine grains (220–600 grain fineness number).
  iii. Grade: Grade represents the hardness of the grinding wheel (soft or hard), which depends upon the strength of the binder and the amount used. As a general rule, soft wheels are used for hard materials, and hard wheels for soft materials. Soft wheels wear fast and thus expose new (fresh) grains to perform the grinding action on hard surfaces.
  iv. Structure: Structure could be open or closed (dense), depending upon the spacing between grains and bond material.
  v. Bond: Bonding could be of five types: V-vitrified, B-resinoid, S-silicate, R-rubber, and E-shellac.

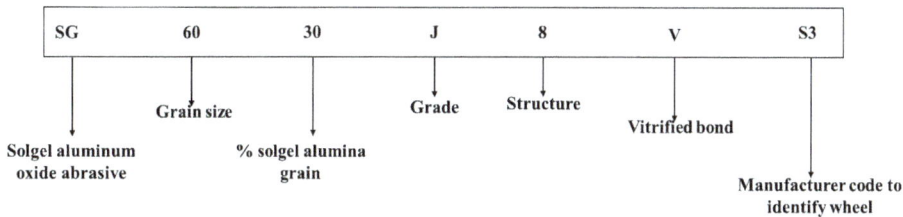

SG	60	30	J	8	V	S3

Grain size — Grade — Structure — Vitrified bond
Solgel aluminum oxide abrasive — % solgel alumina grain — Manufacturer code to identify wheel

Figure 3.14: Alphanumeric systems for grinding wheel specifications.

- **Types of grinding wheels used**
  i. Aluminum oxide wheels: For grinding ferrous materials.
  ii. Silicon carbide wheels: Suitable for non-ferrous materials like brass and aluminum.

iii. Diamond and CBN wheels: Used for high-precision grinding of hard materials.

- **Applications of grinding**
  - Precision flat surface finishing for machine components.
  - Grinding hardened steel, cast iron, and non-ferrous metals.
  - Tool and die manufacturing for high-precision components.
  - Removing material in small increments to achieve close tolerances.
  - Sharpening cutting tools and machine parts.

3.3.5 Drilling operation

Machining round holes is a common operation in the manufacturing industry. Drilling produces a hole in an object by removing material in a circular form by forcing a rotating cutting tool, known as a drill bit. In drilling, a hole is produced by either:
- Forcing a rotating drill tool against a stationary workpiece (drill press).
- Forcing a stationary drill tool against a rotating workpiece (lathe).

1. **Drilling machine**
   The drilling machine is one of the simplest and most accurate machines used in production shops for creating holes in workpieces. The workpiece is securely held and clamped, while the rotating drill bit removes material to form a hole.
2. **Basic components of a drilling machine (as shown in** Figure 3.15)
   i. Table: Supports the workpiece and can be adjusted vertically.
   ii. Belt and pulley: Transmit power from the motor to the spindle for rotation.
   iii. Drill feed handle: Controls the downward movement of the drill bit.
   iv. Spindle head: Houses the spindle, which holds and rotates the drill bit.
   v. Column: Provides structural support to the machine and enables vertical movement of the table.
   vi. Table clamp: Secures the table in the desired position.
   vii. Base: The foundation of the machine, ensuring stability.
   viii. Motor: Powers the spindle and drill bit.

Drill machine, illustrating its key components such as the spindle, chuck, drill bit, and feed mechanism, is used for creating precise holes in various materials.
3. **Operations performed on a drilling machine:** Various operations performed on a drilling machine are shown in Figure 3.16 and discussed below.
   i. Drilling: Drilling produces holes in an object by forcing a rotating drill tool against the work (drill press) or forcing a stationary drill against rotating work (as in a lathe machine). Shown in figure 3.16(b).

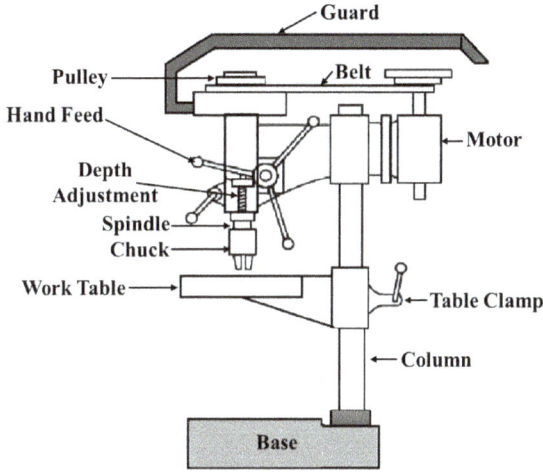

Figure 3.15: Drill machine.

ii. Boring: This is the operation of enlarging a hole that has already been drilled or cast. It produces more accurate (true) holes than drilling. It is an internal turning operation that produces different types of internal surfaces of revolution. See figure 3.16(d).

iii. Reaming: A reamer is a tool that has multiple cutting edges along its circumference, longitudinal to the axis. It is used to remove a small amount of material from the internal surface of a hole. Reaming is performed after drilling and boring operations in order to achieve an accurate diameter and a smooth internal finish for the hole. see figure 3.16(e).

iv. Counter boring: Counter boring is done to enlarge the top portion of a hole to accommodate the head of a fastener. Counter boring is performed after drilling, which provides space for the head of the nut to sit below the surface of the workpiece. The boring tool has a pilot at the end, which guides the tool straight into the drilled hole. See figure 3.16(a).

v. Countersinking: Countersinking is done to create a conical-shaped hole. This operation is performed after drilling to provide space for fasteners with conical-shaped heads to sit flush with the workpiece. Various angles are available for countersinks; some common angles are 60°, 90°, 100°, and 120°. See figure 3.16(f).

vi. Tapping: Tapping is a process of creating internal threads in a drilled hole. Tapping can be performed manually or by machine. Through the tapping process, threads can be cut to the required depth inside the hole. see figure 3.16(c).

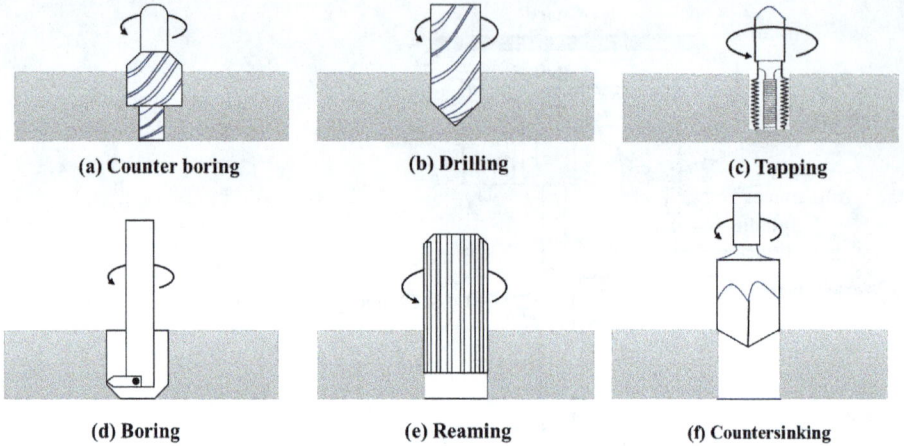

(a) Counter boring	**(b) Drilling**	**(c) Tapping**
(d) Boring	**(e) Reaming**	**(f) Countersinking**

Figure 3.16: Various drilling operations.

3.4 Mechanics of metal cutting

3.4.1 Mechanics of chip formation

During machining operations, metal is removed in the form of chips, which are formed by the plastic deformation of metal through the shear process. The mechanics of chip formation are illustrated below in models. There are two models that exist in this area: the shear plane area model and the primary shear zone model, as shown in Figure 3.17. In these models, the cutting tool moves along the workpiece with certain velocity and depth of cut. A chip is formed just ahead of the tool. The formation of the chip is due to the continuous shearing of the material along the shear plane.

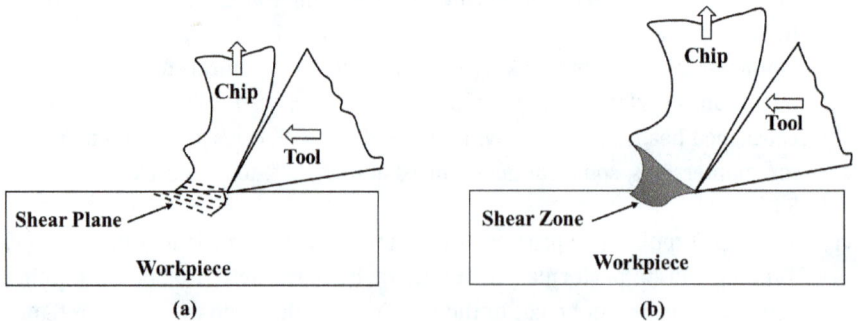

Figure 3.17: (a) Shear plane, (b) shear zone during chip formation.

1. **Types of chips**

   During metal cutting, the type of chip produced influences the surface finish of the workpiece and also affects the overall machining operation. When chips produced during cutting are observed at the microscopic level. Generally, there are three types of chips are seen, as shown in Figure 3.18.

   i. Discontinuous chips (plastic deformation with rupture): When the cutting speed is very high or the rake angle is high, continuous chips are formed. Plastic deformation of the material takes place during the cutting action in the shear zone. These types of chips generally form during the machining of ductile materials.

   ii. Continuous chips (plastic deformation without rupture): When machining brittle materials, they do not undergo high shear strain during cutting, so discontinuous chips are formed. These chips consist of segments that may be loosely attached to each other. Discontinuous chips are also formed at very high or very low cutting speeds.

   iii. Continuous chips with built-up-edge (BUE): They are formed at the tip of the tool during cutting. During cutting, a layer of material from the workpiece is gradually deposited on the tool.

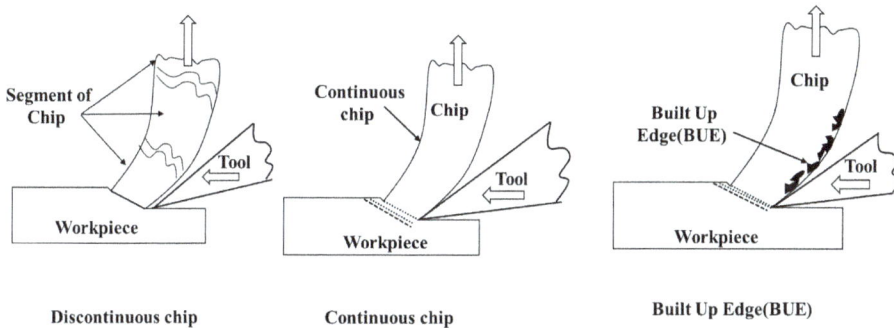

Figure 3.18: Types of chips.

3.4.2 Tool wear

During the cutting operation, the tool is subjected to high cutting forces, elevated temperatures, and rubbing. These are the conditions to wear out the tool. Tool wear is a very important aspect of machining operations because it directly affects the quality of the machined surface and the economics of machining.

1. **Factors involved in tool failure**: There are various factors involved in it; some of them are:

   i. Material of the workpiece and tool (the tool material must be harder than the workpiece material)

    ii. Physical, chemical, and mechanical properties of tool and workpiece materials

    iii. Tool geometry

    iv. Cooling medium

    v. Cutting speed, depth of cut, and feed

– **These are the factors that directly affect the tool life or the failure of the tool.** Broadly, tool failures are illustrated in Figure 3.19:

    i. Gradual microscopic wear

    ii. Mechanical impact causes failure of the tool

    iii. Thermal stress causes failure

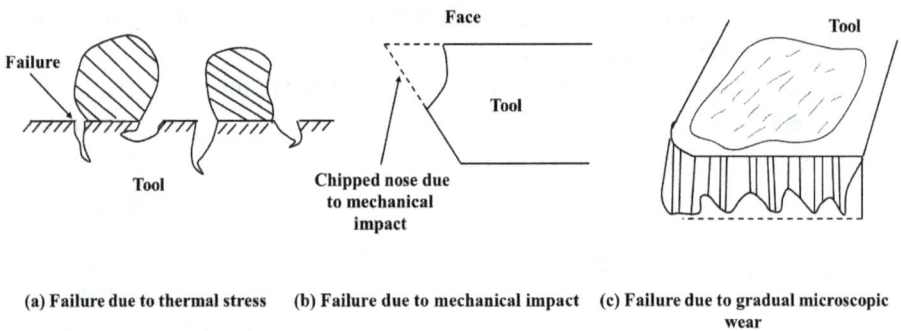

(a) Failure due to thermal stress   (b) Failure due to mechanical impact   (c) Failure due to gradual microscopic wear

Figure 3.19: Types of tool failures.

2. **Indication of tool failure**

    i. When the tool is completely failed or has broken down.

    ii. Due to excessive rubbing, the level of noise increases.

    iii. When consumption power and feed increased, it shows the tool is about to fail.

    iv. Weight loss of the tool tip due to wear indicates that the tool has failed.

    v. Excessive nose wear causing significant machining errors.

    vi. Development of large-sized flank-wear land-causing rubbing.

    vii. Large crater wear development.

    viii. When the surface finish is not good.

a) **Flank wear:** When the tool slides along the machined surface, it causes abrasive wear. Due to this abrasive action, the temperature rises, which adversely affects the tool material. Flank wear, as shown in Figure 3.20, occurs on the flank face of the tool.

F.W. Taylor published his extensive study regarding the relationship between tool wear and cutting various steels:

$$VT^n = C$$

where V is the cutting speed, T is the time in minutes it takes to develop flank wear land, n is the exponent that depends on cutting conditions, and C is a constant.

b) **Crater wear:** The factors that affect flank wear also influence crater wear, as shown in Figure 3.20. This type of wear occurs on the top face of the tool at a short distance from the cutting edge. The most important factors causing this wear are temperature and the degree of chemical affinity between the tool and the workpiece. When the chip rubs against the top face of the tool, the temperature at the interface rises, and the chip becomes alloyed with the tool face. A layer of alloy is formed on the tool face, which is carried away by the chip due to the high pressure exerted by the chip on the face of the tool. This wear occurs due to diffusion, abrasion, oxidation, and adhesion.

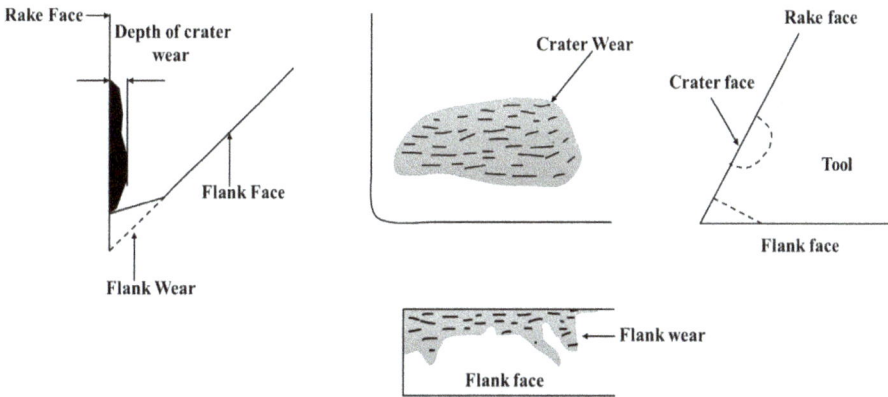

Figure 3.20: Flank and crater wear.

3.4.3 Cutting fluid and lubricants

In every cutting operation, due to high friction, high temperatures are generated. These high temperatures cause tool failure or reduce tool life. Cutting action can be improved by the use of solids, liquids, emulsions, or gases in the cutting process.

- **Advantages of cutting fluid**
  - Cutting fluid reduces friction, which causes a reduction in temperature.
  - It improves tool life and surface finish.
  - It reduces power consumption and cutting force.
  - Due to the reduction in temperature, distortion in the cutting zone is also reduced.

- Cutting fluid provides protection to newly machined surfaces from environmental effects like corrosion.
- Move the chip out of the workpiece if cutting fluid is applied with a certain pressure.
- Cutting fluid protects the welding of chips to the tool.

3.5 Unconventional machining

In the following paragraphs, the important non-conventional machining processes that are used in modern industries have been discussed. The discussion also includes the working principles, applications, limitations, and benefits of using these processes. Each process includes a summary of the discussion at the end for quick revision. Among the machining processes, EDM, ECM, LBM, EBM, USM, AJM, and WJM have been specially chosen for the present discussion.

3.5.1 Electro-discharge machining (EDM)

The process uses a thermo-electric source of energy that permits the machining of difficult-to-machine materials through controlled erosion, achieved by a series of sparks submerged in a dielectric fluid medium (kerosene or mineral oil). This dielectric fluid is fed to the workpiece under pressure. It cools the tool, washes away particles of eroded metal from the workpiece, and maintains a uniform resistance to the flow of current. The heat of the spark raises the temperature to 10,000–12,000 °C. The workpiece is made the ANODE. Since machining can be carried out in the hardened state of the work material, subsequent heat treatment after machining is eliminated, thereby reducing the possibility of cracking.

Figure 3.21 depicts the schematic arrangement of electric discharge machining (EDM). A capacitor, connected in parallel with the electrode and workpiece, is charged by a DC power source through a resistor. As the capacitor energizes, its voltage rises until it reaches a level sufficient to break down the dielectric fluid between the electrode and workpiece. The gap distance (ranging from 0.025 to 0.076 mm) is electronically regulated by a servo mechanism to maintain a stable potential. Sparking occurs at the point where the gap is at its smallest.

Most EDM machines include a frequency selector, which controls the number of sparks per second between the electrode and the workpiece. The time interval between sparks is 10 to 30 microseconds, with a current density of 15–500 A/mm^2. Common electrode materials include tungsten carbide, copper-tungsten, and graphite. To minimize the risk of fire, a dielectric fluid level of 50 mm is maintained above the work surface. Additionally, a spark gap of approximately 0.025 to 0.05 mm is controlled by the servomotor.

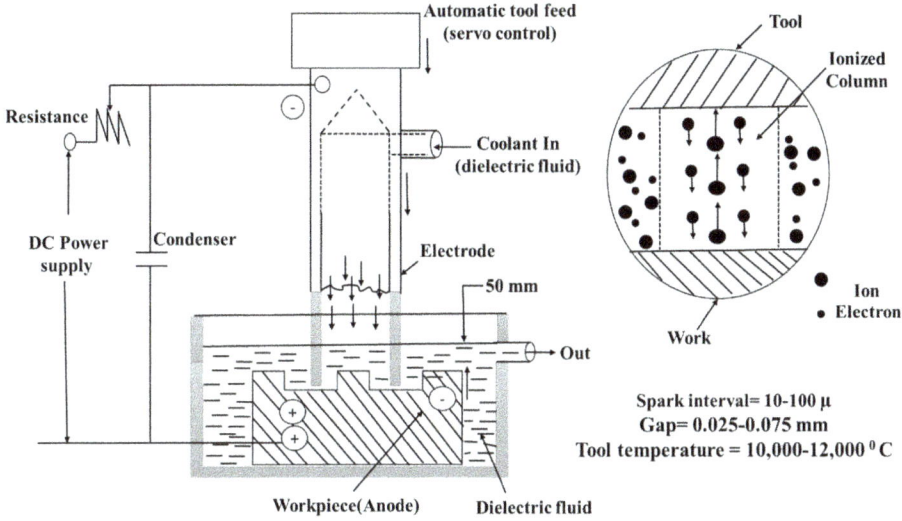

Figure 3.21: Basic schematic of the electric charge mechanism.

Table 3.3 presents a comprehensive summary of electro-discharge machining (EDM), detailing essential parameters such as the metal removal mechanism, dielectric medium, tool materials, tool–workpiece gap, metal removal rate, and specific power consumption. It also outlines critical factors like voltage and dielectric circulation, as well as typical applications and limitations of the process, making it a valuable reference for understanding EDM operations.

Table 3.3: Summary of electro-discharge machining (EDM).

Parameter	Details
Metal removal mechanism	Melting, evaporation/combined with cavitation
Medium	Dielectric fluid
Tool material	Copper, brass, copper-tungsten alloy, silver-tungsten alloy, and graphite
Tool and workpiece gap	10–125 µm
Metal removal rate	5,000 mm^3/min
Ratio of MRR/tool wear rate	0.1 to 10
Specific power consumption	1.8 W/mm^3/min
Critical parameters	Voltage, capacitor, spark plug, dielectric circulation

Table 3.3 (continued)

Parameter	Details
Application	Can cut all conducting metals and alloys, blind holes, complex shapes, micro-holes (for nozzles), non-circular holes, and narrow slots
Limitation	High specific energy consumption at high MRRs; the finish is rough, and non-conductors cannot be machined

3.5.2 Electrochemical machining (ECM)

Electrochemical machining (ECM), as shown in Figure 3.22, is an advanced, non-traditional machining process that removes material from a workpiece through electrochemical dissolution rather than mechanical cutting. It is especially effective for machining hard materials and complex geometries with high precision.

In electrical conductors, electricity flows through the movement of free electrons, whereas in an electrolyte, electrical conduction occurs via the movement of ions. During ECM, material is dissolved from the anode (workpiece), with the dissolution rate being higher where the gap between the tool and workpiece is smaller, as the current is inversely proportional to the gap distance. This process is used when the work material has very low machinability or when the shape of the machined component is complex.

– **Working principle**
  – ECM operates on the principle of electrolysis.
  – A DC power supply is used to create an electrolytic cell between the workpiece (anode, positive) and the tool (cathode, negative).
  – An electrolyte solution (usually sodium chloride or sodium nitrate) flows between the tool and the workpiece.
  – When electricity passes through the electrolyte, metal ions from the workpiece dissolve, removing material without direct contact.

– **Components of ECM**
  i. Power supply: Provides DC voltage (typically 5–25 V).
  ii. Electrolyte system: Circulates and filters the electrolyte to remove dissolved materials.
  iii. Tool (cathode): Shaped according to the desired cavity or profile.
  iv. Workpiece (anode): Material to be machined.
  v. Pump and nozzle system: Maintains a continuous electrolyte flow.
  vi. Filtration system: Removes metal hydroxides from the electrolyte.
– **Electrochemistry of ECM:** ECM is based on Faraday's laws of electrolysis, where material removal occurs through controlled electrochemical dissolution in an

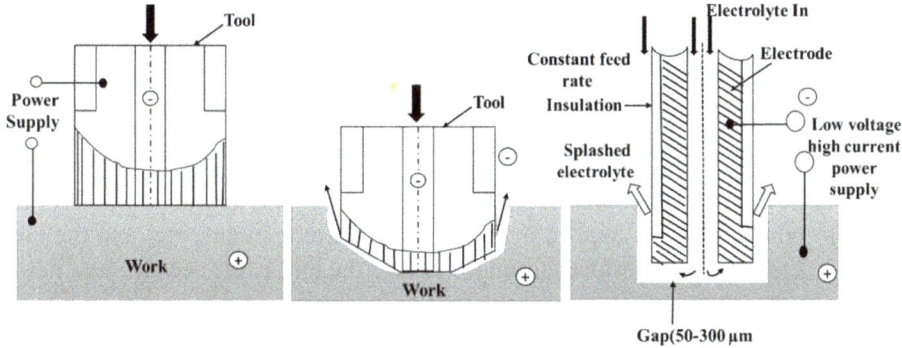

Figure 3.22: Electrochemical machining.

electrolyte solution. The process is the reverse of electroplating, as shown in Figure 3.23.

Construction of ECM chemistry is mentioned as follows:
- The workpiece (anode) is connected to the positive terminal of a DC power source.
- The tool (cathode) is connected to the negative terminal and does not wear out.
- An electrolyte (e.g., NaCl or $NaNO_3$ solution) flows between the anode and cathode, conducting electricity and removing dissolved material.
- When a DC voltage (5–25 V) is applied, metal ions dissolve from the anode into the electrolyte.
- Chemical reactions in ECM are shown as:

At the anode (workpiece): Metal dissolution for a metal (M) dissolving into a solution:

$$M \rightarrow M^{n+} + ne^-$$

Example for **iron (Fe):** $Fe \rightarrow Fe^{2+} + 2e^-$
At the cathode (Tool): Hydrogen evolution $2H_2O + 2e^- \rightarrow H_2 + 2OH^-$
This reaction **produces hydrogen gas** at the tool's surface.

Electrolyte reaction: The electrolyte helps in carrying away the dissolved metal ions. For a **sodium chloride (NaCl) solution**, the reaction is:

$$Fe^{2+} + 2Cl^- \rightarrow FeCl_2$$

This prevents the passivation of the workpiece.
- **Let us take an example** of the machining of low-carbon steel, a ferrous alloy primarily composed of iron, which can be machined using ECM. A neutral salt solution of sodium chloride (NaCl) is commonly used as the electrolyte, as shown in

Figure 3.24. When a potential difference is applied, the electrolyte and water undergo ionic dissociation as follows:

$$NaCl \leftrightarrow Na^+ + Cl^-$$

$$H_2O \; H \leftrightarrow^+ + (OH)^-$$

– **Electrochemical reactions in ECM**
a) When voltage is applied between the workpiece (anode, +ve) and the tool (cathode, −ve), positive ions move toward the tool, while negative ions move toward the workpiece.
b) At the cathode (tool), hydrogen ions accept electrons and form hydrogen gas:

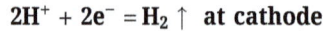

$$2H^+ + 2e^- = H_2 \uparrow \quad \text{at cathode}$$

c) At the anode (workpiece), iron atoms lose electrons and dissolve into the electrolyte:

$$Fe = Fe^{++} + 2e^-$$

d) Within the electrolyte, iron ions react with chloride ions to form iron chloride ($FeCl_2$), while sodium ions combine with hydroxyl ions to form sodium hydroxide (NaOH):

$$Na^+ + OH^- = NaOH$$

e) In practice, iron hydroxide ($Fe(OH)_2$) and iron chloride ($FeCl_2$) precipitate as sludge. As machining progresses, the workpiece material is gradually removed and converted into sludge, ensuring a stress-free and burr-free surface.

Additionally, no material is deposited on the tool, as only hydrogen gas evolves at the cathode. Since the removal occurs at an atomic level, the machined surface is exceptionally smooth and free from mechanical stresses.

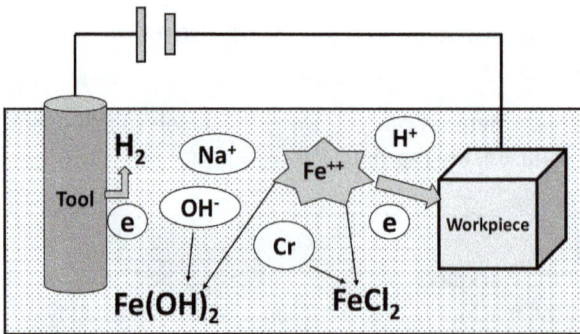

Figure 3.23: Schematic representation of electrochemical reactions.

– **Factors affecting ECM chemistry**
  i.   Electrolyte composition: Common electrolytes include NaCl, $NaNO_3$, and KCl, which enhance conductivity and prevent passivation.
  ii. Electrolyte flow rate: It must be sufficient to remove dissolved metal ions and maintain process stability.
  iii. Current density: Higher current leads to faster material removal but may result in increased surface roughness.
  iv. Inter-electrode gap: Maintained at 0.025–0.5 mm to control the dissolution rate.

– **Modeling of material removal rate (MRR) in ECM:** The MRR in ECM depends on **Faraday's laws of electrolysis**, which relate the amount of material removed to the electrical charge passed through the electrolyte. The MRR can be mathematically modeled as follows:
  a) **Faraday's first law of electrolysis**

The mass of material removed (m) is directly proportional to the total charge (Q) passed:

$$m \propto Q$$

$$m = ZQ/F$$

$$(Z = M/nF)$$

When a metallic body (electrode A) is submerged in an electrolyte, as shown in Figure 3.24, the metallic atoms leave the body and become atoms. The process is continuous, and equilibrium is maintained. A potential difference exists between the metallic body and an adjacent point in the electrolyte.

– **Current-based material removal rate**

Since Q = It (Charge = Current × Time), substituting into the above equation:

$$m = MIt/nF$$

The **MRR**, in terms of **volume**, can be expressed as:

$$MRR = MI/nF\rho$$

where m = Weight (in grams) of material dissolved/deposited
  Q = Amount of charge passed
  Z = Electrochemical equivalent of the material
  F = Constant of proportionality (Faraday's constant = 96,500 Coulombs)
  n = Number of electrons involved in ionization
  M = Atomic mass of the workpiece material (g/mol)
  I = Current (A)

t = Machining time (s)

ρ = Density of the material (g/cm^3)

MRR = Material removal rate (cm^3/s)

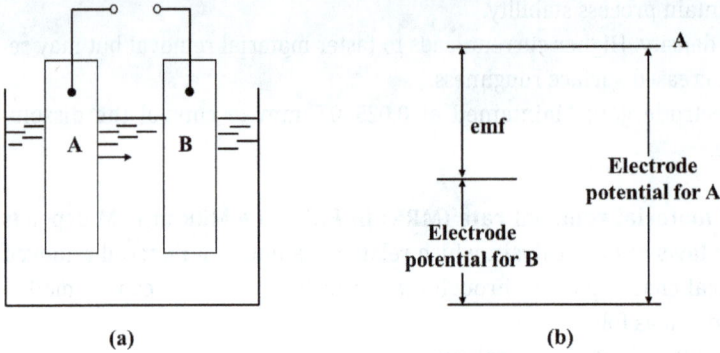

Figure 3.24: Electrode dissolution and electrode potential. (a) Material dissolution from A. (b) Cell EMF and deposition at B.

- **Factors affecting MRR in ECM**
  i.   Applied current (I): Higher current increases the material removal rate.
  ii. Inter-electrode gap (d): A smaller gap increases current density, thereby enhancing MRR.
  iii. Electrolyte conductivity (σ): Higher conductivity improves ion transport and MRR.
  iv. Atomic mass (M) and valency (n): Materials with lower valency dissolve faster.
  v.   Density of workpiece (ρ): Denser materials require more time to machine.

- **Functions of electrolyte for ECM**
  i.   Complete electrical circuit between electrode and work piece.
  ii. Allow electrochemical reactions to be produced efficiently.
  iii. Carry away the products of the reaction and the heat generated.
  iv. Carry away the heat generated during the process.

- **Essential properties of a good electrolyte are:**
  i.   High thermal conductivity.
  ii. Low viscosity and high specific heat.
  iii. Non-passivating nature.
  iv. Chemically stable.
  v.   Non-corrosive and non-toxic.
  vi. Economical and easily available.
  vii. High electrical conductivity.

- **Some typical applications of ECM are:**
  i.   Drilling long and slender holes.

   ii.   Generation of three-dimensional surfaces.

   iii. Die sinking, profiling of odd shapes.

   iv. Multiple-hole drilling.

   v.   Trepanning.

   vi. Broaching.

   vii. Deburring.

   viii. Grinding and polishing.

– **Advantages of ECM:**

   i.   Tool wear is negligible.

   ii.   Machining is done at low de voltage (5–15 Vdc) and high currents (10,000 A to 30,000 A), with MRRs of 550 mm^3/s.

   iii. Surface finish is extremely good (0.1 μm) R_a with a tolerance of the order of 0.005 mm.

   iv. Any metal with good electrical conductivity can be given any complicated profile by choosing a properly shaped tool.

   v.   Chemical and mechanical properties of work remain unaffected.

   vi. Toughness or brittleness of materials has no effect on the machining process.

   vii. No thermal damage occurs to the workpiece due to the low temperature developed during normal machining.

– **Limitations of ECM**

   i.   High consumption of electrical energy.

   ii.   The process is comparatively slow.

   iii. Work material should be a good conductor of electricity.

   iv. At low current densities (5 A/cm^2), oxygen evolves at the anode, which consumes the major part of the energy supplied, resulting in low MRR. In order to obtain high current efficiently, higher current densities of the order of 200 A/cm^2 are used.

   v.   Sharp internal corners cannot be achieved.

   vi. Post-machining cleaning is a must to protect the work from corrosion.

   vii. Polarized ionic layers may build up at either electrode, causing large voltage drops at the surfaces. This effect can be reduced by increasing the flow of clean electrolyte in the cathode-anode gap, which is very small (0.025 to 0.125 mm). A high flow rate of electrolyte will require high pressure (2.5 N/mm^2) to maintain a flow velocity of 30 m/s. The electrolyte temperature in the tank is maintained at 35–65 °C.

The key characteristics of ECM process has been summarized in Table 3.4, including its electrolysis-based metal removal mechanism, tool wear rate, feed rate, surface finish, and electrolyte properties.

Table 3.4: Summary of ECM characteristics.

Parameter	Details
Mechanics of metal removal	Electrolysis
Medium	Conducting electrolyte
MRR/tool wear rate	Infinite
Working gap	0.1 to 2 mm
Overcut	0.1 to 3 mm
Feed rate	0.5 to 15 mm/min
Electrode material	Copper, brass, bronze
Surface roughness	R_a 0.2 to 1.5 µm or 15 microns
Tolerance	0.025 mm (2D) and 0.050 mm (3D)
Maximum MRR	15×10^3 mm^3/min
Critical parameter	Voltage, current feed, electrolyte conductivity
Materials application	All conducting metals and alloys
Shape application cutting	Blind complex cavities, curved surfaces, and large through-cutting
Electrolyte a) Material b) Temperature c) Flow rate d) Velocity e) Dilution f) Pressure	 NaCl, NaNO$_3$, and proprietary mixtures 20–50 °C 16 to 20 LPM 1,500 to 3,000 m/min 100 to 500 g/l 0.5 to 20 bar **Inlet pressure**: 2,200 kPa **Outlet pressure**: 300 kPa
Power supply a) Type b) Voltage c) Current d) Current density e) Specific power consumption (typical)	 Direct current 2 to 35 V 50 to 40,000 A 0.1 to 5 A/mm^2 7 W/mm^3/min

Apart from machining, the electrochemical technique is also used for grinding and deburring. The process is discussed below.

1. **Electrochemical grinding (ECG)**: Electrochemical grinding (ECG) is a hybrid machining process that combines ECM and conventional grinding to remove material from a workpiece with high precision and minimal tool wear. It is a process that removes electrically conductive material by grinding with a negatively charged abrasive

grinding wheel and a positively charged work piece. Material removed from the workpiece is carried away by the electrolyte fluid present. Electrochemical grinding and electrochemical machining are principally similar to each other; however, a grinding wheel is used instead of a tool, which is a replica of the contour to be obtained on the work piece.

- **Working principle**
    a) A **rotating conductive grinding wheel** (cathode) is used, with an **electrolyte** flowing between the wheel and the workpiece (anode).
    b) **Electrochemical dissolution** removes about **90%** of the material, while the abrasive action of the wheel removes the remaining **10%**.
    c) **No direct contact** between the wheel and the workpiece reduces heat generation and tool wear.
- **Difference between ECM and ECG**
  ECG is similar to ECM but uses a contoured, conductive grinding wheel instead of a tool shaped like the contour of the workpiece.
- **Advantages**
    - **Burr-free and stress-free machining.**
    - **Higher MRR** compared to conventional grinding.
    - **No thermal damage** to the workpiece.
    - Can machine **hard and tough materials**, including superalloys and hardened steels.
- **Applications**
    - Aerospace and medical components.
    - Machining **turbine blades, hypodermic needles, and surgical instruments**.
    - **Sharpening cutting tools** with minimal heat effects.

2. **Electrochemical deburring (ECD):** Electrochemical deburring (ECD) is a **non-traditional deburring process** that removes unwanted burrs and sharp edges from metallic components using electrochemical dissolution. It is a localized deburring process that uses electrical energy to remove burrs in a specific location, as opposed to TD, which provides general burr removal.

- **Working principle**
    - The workpiece (anode), charged positively, and a shaped tool (cathode), charged negatively, are submerged in an electrolyte solution.
    - A DC voltage is applied, causing controlled material removal from burrs and sharp edges.
    - The tool is positioned near the burrs to selectively remove excess material without affecting the main surface.
- **Advantages:**
    - Selective burr removal without affecting critical dimensions.
    - No mechanical stress or heat-affected zones.

- Ideal for complex and delicate parts.
- Can be used for internal deburring, where mechanical methods are difficult.
- **Applications**
  - Deburring precision parts in the aerospace, automotive, and medical industries.
  - Removing burrs from gears, hydraulic components, and fuel injectors.
  - Finishing micro-sized features in electronic and medical devices.

3. **Electrochemical honing (ECH):** Electrochemical honing (ECH) is a hybrid machining process that combines ECM and mechanical **honing** to improve surface finish and dimensional accuracy, particularly in cylindrical components.
- **Working principle**
  - The workpiece (anode, +ve) and a honing tool (cathode, −ve) are submerged in an electrolyte solution (e.g., NaCl or NaNO$_3$).
  - A DC power supply initiates electrochemical dissolution, removing most of the material.
  - Simultaneously, honing stones apply abrasive action to improve the surface finish.
  - The electrolyte flow removes dissolved metal ions, preventing tool wear.
- **Advantages**
  - Higher MRR compared to conventional honing.
  - Superior surface finish and improved roundness.
  - No tool wear occurs due to electrochemical dissolution.
  - Stress-free and burr-free machining.
- **Applications**
  - Finishing internal bores in engine cylinders, hydraulic cylinders, and aerospace components.
  - Honing precision tubes in the medical and nuclear industries.
  - Used for hard and tough materials, such as titanium and nickel alloys.

3.5.3 Electron beam machining (EBM)

EBM is a thermal material removal process in which a high-speed electron beam impinges on the workpiece, transferring kinetic energy and generating intense localized heating. This heat causes the material to melt or vaporize, enabling precise machining. A schematic view of the electron beam machining set-up is shown in Figure 3.25.
- **Working principle of EBM**
  - A beam of electrons is emitted from an electron gun, which is basically a triode consisting of:
    - A cathode, which is a hot tungsten filament (2,500 °C) emits high negative potential electrons.

  - A grid cup, negatively biased with respect to the filament.
  - An anode, which is heated at ground potential and through which high-velocity electrons pass.
 - High-speed electrons (accelerated up to 228,478 km/s) impact the work surface.
 - The power density can reach 6,500 billion W/mm^2, which is enough to vaporize any material instantly.
 - The beam is focused to a small diameter (10–200 μm), allowing for precise drilling and cutting.
 - The process occurs in a vacuum (~10^{-5} mm Hg) to prevent electron collisions with air molecules.
- **EBM setup components:**
  i.   Tungsten cathode: Emits electrons.
  ii.   Anode: Accelerates electrons using a voltage of 20–150 kV.
  iii. Grid cup: Shapes the electron beam.
  iv.   Electromagnetic lenses: Focus the beam on the workpiece.
  v.   Deflecting coils: Control beam movement for complex machining.

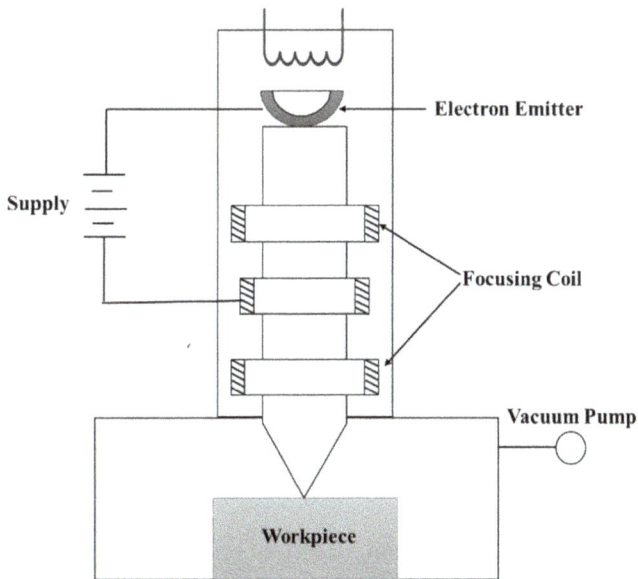

Figure 3.25: Schematic view of the electron beam machining setup.

 - **General cutting process by electron beam**
   - Relative movement between the beam and the workpiece is possible to secure.
     - Moving the workpiece on the compound desk: macromachining, trepanation of larger holes.

- Deflection of electron beam by deflection reel: micromachining.
- Electron lithography: engraving of a mask on the workpiece surface before chemical processing (electronic component).
- **Analysis in EBM**
  - MRR/machining rate

$$Q = A \times V$$

where Q = Material removal rate (mm³/s)
  A = Area of the hole or slot (mm²)
  V = Cutting speed (mm/s)
- The velocity of electron impingement,

$$V_s = 600 \sqrt{E_s}$$

where V_s = Velocity of electrons striking the workpiece (km/s)
  E_s = Voltage of the electric field (V)
- The power of electron beam

$$P_b = E_s I_b \quad W$$

where P_b = Power of the electron beam (W)
  I_b = Beam current (A)
  E_s = Beam accelerating voltage (V)
- The electron beam pressure

$$F_b = 0.34 \times I_b \sqrt{E_s} \quad dyne/cm^2$$

where F_b = Electron beam pressure (dyne/cm²)
  I_b = Beam current (A)
  E_s = Beam accelerating voltage (V)
- The thermal velocity acquired by an electron

$$V_a = \sqrt{(2k\theta/M_a)} \quad m/s$$

where V_a = Thermal velocity of an electron (m/s)
  k = Boltzmann's constant = 1.38×10^{-23} J/K/atom
  θ = temperature raised through electron bombardment, K
  M_a = mass of one atom of the workpiece, kg
- **Advantages of EBM**
  - Very small holes can be machined with high accuracy in all types of materials.
  - Holes of different shapes and sizes can be machined.
  - Little physical and metallurgical damage to the workpiece.

- No tool wear occurs as there is no mechanical contact between the tool and the work piece.
- It provides a concentrated source of heating.
- **Disadvantages of EBM**
  - Equipment and set-up cost is considerably high.
  - Low metal removal rates.
  - Necessitates a skilled operator.
  - Used only for small cuts.
  - Vacuum requirements limit the size of the work that can be handled.
- **Applications**
  - Drilling of holes in pressure differential devices used in aircraft engines, nuclear reactors, etc.
  - Manufacture of small cross-sectional wire drawing dies.
  - Scribing of thin films.
  - Drilling holes in ruby and diamond crystals.
  - Micro-welding and drilling.
  - Drilling of apertures in electron microscopes.
  - Removing small broken tops from holes.

The essential characteristics of EBM are outlined in Table 3.5, which includes parameters such as the metal removal mechanism (melting and vaporization), operating medium (vacuum), accelerating voltage, material applicability, and typical shape applications like fine hole drilling and contour cutting.

Table 3.5: Summary of ECM characteristics.

Parameters	Details
Mechanics of metal removal	Melting, vaporization
Medium	Vacuum
Accelerating voltage	20–150 kV
Tool	Beam of electrons moving at a very high velocity (228,478 km/s)
Max. MRR	10 mm^3/min (0.01 mg/s normal)
Specific power consumption (typical)	450 W/mm^3-mm
Critical parameters	Accelerating voltage, beam current, beam diameter, work speed, melting temperature
Material application	All materials
Shape applications	Drilling fine holes (0.05 mm dia), cutting contours in sheets, and cutting narrow slots

Table 3.5 (continued)

Parameters	Details
Limitations	Very high specific energy consumption
	Necessity of vacuum, expensive machine
	Large-sized jobs are limited to the size of the vacuum chamber
	Taper of 1–20 is observed in thicker plates
	The process results in X-ray emissions; the work area must be shielded to absorb radiation

3.5.4 Laser beam machining (LBM)

LASER stands for **"Light Amplification by Stimulated Emission of Radiation."** It is an intense, monochromatic, and coherent beam of electromagnetic radiation. The concept of laser technology is based on the principle of stimulated emission, wherein an external energy source excites electrons in an atom, causing them to release additional light of the same frequency and wavelength as the original source.

– Key characteristics of laser light:
  i. Monochromatic: Laser light consists of a single wavelength, unlike white light, which contains multiple colors.
  ii. Coherent: The light waves travel in phase, allowing for higher intensity and better focus compared to ordinary light.
  iii. Collimated: Laser beams are highly directional, with minimal divergence (typically less than 1–2 mrad), ensuring energy concentration over long distances.
  iv. High power density: When focused using a lens, laser beams can achieve extremely high power densities, making them effective for industrial applications.
– Comparison of light bulb and laser power density:
  – A 100 W light bulb at a distance of 1 meter provides an intensity of 0.0008 W/cm^2.
  – A 100 W laser with a 1 cm diameter beam at the same distance produces 127 W/cm^2.
  – When focused on a 0.127 mm spot, the power density can exceed 800,000 W/cm^2.
– Power density requirements for material processing:
  – Surface heating: 1.5×10^2 to 1.5×10^3 W/cm^2
  – Melting (welding): 1.5×10^4 to 1.5×10^6 W/cm^2
  – Cutting and drilling: 1.5×10^6 to 1.5×10^8 W/cm^2
  – Sublimation (solid to vapor transition): More than 1.5×10^8 W/cm^2
– Principle of laser operation: Stimulated emission and population inversion

a) **Absorption and energy transition**
 – The core concept is represented in Figure 3.26(a) as the transition by absorption and spontaneous emission between two energy levels of an atom, whereas (b) illustrates the energy diagram for an atom showing level-to-level transitions.
 – When an atom at energy level q absorbs a photon of the correct frequency, it moves to a higher energy level p.
 – Conversely, when an atom transitions from a higher energy level, p, to a lower-level, q, it emits energy in the form of a photon.
 – This energy emission occurs in two ways:
    – Spontaneous emission: Occurs naturally, independently of light intensity.
    – Stimulated emission: A photon triggers an excited atom to release another identical photon, thereby amplifying the light.

ii. **Stimulated emission and laser operation**
 – Each horizontal energy level represents the allowed energy states of an atom.
 – If an atom is excited to a higher energy level, E2, and then returns to its ground state, E0, it releases a photon.
 – When this photon interacts with another excited atom at E2, it stimulates the emission of an additional identical photon (same wavelength, phase, direction, and energy).
 – This chain reaction produces a coherent beam of identical photons, forming the basis for laser generation.
 – Population inversion.
 – For a laser to function effectively, more atoms must be in the higher-energy state than in the lower-energy state.
 – This condition is known as population inversion, which is achieved by supplying external energy (pumping) to the lasing material.

iii. **Feedback mechanism**
 – A feedback system captures and redirects some of the emitted coherent photons back into the active medium, further stimulating emission.
 – A small portion of the photons exit as a laser beam, while the rest maintain the amplification process.
 – This laser output is used in industrial, medical, and military applications.

1. **Laser beam machine:** A laser beam is a highly coherent, monochromatic beam of electromagnetic radiation with wavelengths ranging from $0.21\ \mu m$ to $11\ \mu m$. Common laser types and their wavelengths include:
 – Ruby laser: $0.7\ \mu m$
 – Nd:YAG laser: $1\ \mu m$
 – CO laser: $2.7\ \mu m$
 – CO_2 laser: $10.6\ \mu m$

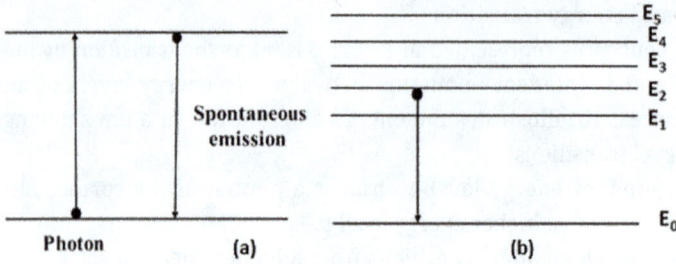

Figure 3.26: (a) Transition by absorption and spontaneous emission between two energy levels of an atom, and (b) energy diagram for an atom showing level-to-level transitions.

For machining applications, the usable wavelength range is typically 0.4–0.6 μm. The laser beam can achieve extremely high power densities of up to 10^{12} W/mm^2 when focused on a small spot.

– **Laser types used in machining:**
   i. Pulsed ruby laser: Commonly used for micro-drilling and fine cutting.
   ii. Continuous CO_2-N_2 laser: Suitable for machining applications requiring a continuous energy input.
– **Components of laser beam machining (LBM)**

LBM consists of several essential components, as shown in Figures 3.27 and 3.28, that generate, direct, and focus the laser beam for precise material removal.
   i.   Laser source (active medium)
   ii.   Energy supply (pump source)
   iii. Optical resonator (mirrors and cavity)
   iv.   Beam focusing system (lens and beam delivery system) for a fine spot (~10–200 μm).
   v.   Assist gas system: Oxygen, argon, helium, or nitrogen
   vi.   CNC control system
   vii. Cooling system
   viii. Exhaust and safety system
– **Working principle:**
   – The laser source generates a high-energy beam, which is reflected within the optical cavity.
   – The beam exits through a partially reflective mirror and is focused onto the workpiece.
   – Intense heat causes localized melting, vaporization, or sublimation, enabling precise material removal.
   – **Electrical discharge excites gas atoms → Excited atoms release photons → Photons stimulate further emissions → Optical resonator amplifies light → Laser beam exits through a partially reflecting mirror.**

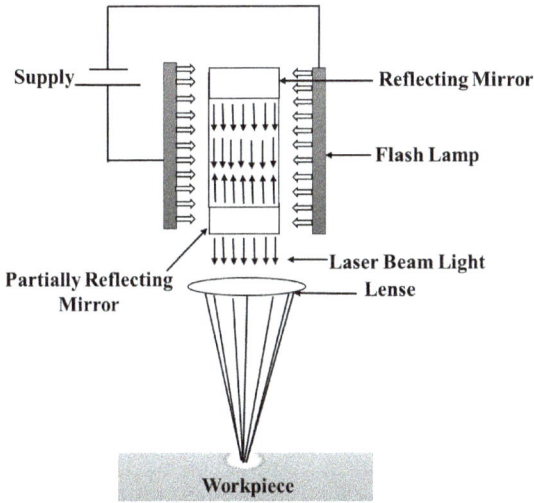

Figure 3.27: Schematic diagram of LBM setup.

Schematic diagram of laser beam machining (LBM) setup, illustrating key components such as the laser source, focusing lens, workpiece, and assist gas for precise material removal.

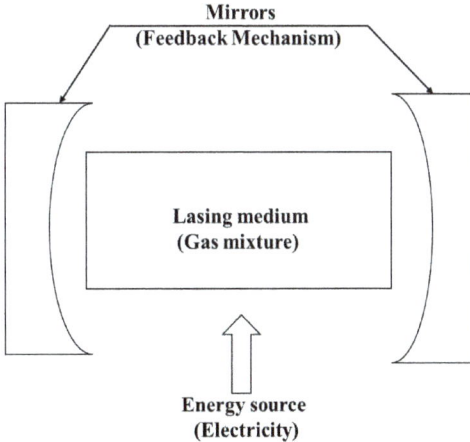

Figure 3.28: Components of a gas laser.

Components of a gas laser illustrate key parts such as the discharge tube, mirrors, electrodes, and gas medium, which are essential for laser generation and amplification.

- **Machining requirements**
    - The required laser beam power density for machining ranges from 1.5×10^4 to 1.5×10^8 W/cm^2.
    - Similar to EBM, laser beams are used for micro-hole drilling and narrow-slot cutting.
    - Holes up to 250 μm in diameter can be drilled.
    - For workpiece thickness >0.25 mm, a taper of 0.05 mm may be observed.
- **Operating parameters**
    - The flash lamp frequency varies from one flash every three minutes to 12 flashes per minute.
    - Energy application time on the workpiece is less than 0.002 s.
- **Process characteristic**
    - The process has high capital and operating costs and low machining efficiency (>1%). This makes it in competitive with conventional machining.
    - Machining with a laser reduces the fatigue strength of the machined component compared to conventionally machined parts.
    - As there are no mechanical forces exerted on the work piece, LBM is capable of machining refractory, brittle, hard metallic and non-metallic materials (e.g., cast alloys, tungsten, alumina, and diamond).
    - It can operate through a transparent environment (air, gas, vacuum).
    - It cannot be used for highly conductive or highly reflective materials.
    - Very small diameter holes (230 μm dia) and holes with a large aspect ratio can be machined using this process. For good-quality holes, high peak power and high-power density are recommended. The recommended pulse duration for deep hole drilling is 0.1–2.5 ms.
- **MRR in LBM:** MRR in LBM depends on the power density, interaction time, and thermal properties of the workpiece. It is primarily governed by melting, vaporization, or sublimation.

The MRR can be estimated using:

$$MRR = P/\rho Hm$$

where MRR = Material removal rate (mm^3/s)

P = Laser power (W)

ρ = Density of the material (g/cm^3)

Hm = Specific enthalpy required for phase transformation (J/g)

- **Factors affecting MRR in LBM**
    - Laser power (P): Higher power results in faster material removal.
    - Beam spot size: Smaller spots result in higher energy density and faster machining.
    - Material properties: Higher melting and vaporization temperatures reduce the MRR.

- Pulse duration and frequency: Short pulses with high frequency improve precision and efficiency.
- Assist gas (O_2, N_2, Ar): Helps in material ejection and affects cutting speed.
- **Advantages of laser beam machining**
  - Dissimilar metals can easily be welded.
  - Small heat-affected zone.
  - There is no direct contact between the tools and the workpiece, allowing fragile materials to be machined. Additionally, there is no tool wear.
  - Ease of control.
  - Refractory, brittle, and hard materials cannot be easily worked.
  - Extremely small holes can be machined.
- **Disadvantages of laser beam machining**
  - Overall efficiency is extremely low (10–30%).
  - The process is limited to the machining of thin sheets/plates.
  - Life of a flash lamp is short.
  - The cost of the equipment is very high.
  - Metal removal rate is slow.
  - Machined holes are neither round nor straight.
  - Certain materials like fiber glass, reinforced materials, and phenolics cannot be worked on by laser as they burn, char, and bubble.
- **Applications**
  - Machining.
  - Welding.
  - Heat treatment.
  - Micro-machining.
  - Sheet metal trimming, engraving, and blanking can be done.
  - Determining the distance, velocity, direction, size, and form of distant objects by means of selected radar signals.

Table 3.6 outlines the main characteristics of the LBM process, while Table 3.7 compares LBM with EBM in terms of working principles, precision, and applications.

3.5.5 Ultrasonic machining

The story of USM starts with a few questions, such as: What is the difference between frequency, wavelength, and amplitude?
- Frequency tells us how many waves pass through a point in one second.
- Wavelength tells us the length of these waves.
- Amplitude tells us how large the wave is.

Table 3.6: Summary of LBM process characteristics.

Parameter	Details
Metal removal mechanism	Melting and vaporization. The typical output energy is 20 J with a pulse duration of 5 ms
Medium	Normal atmosphere
Tool	High-power laser beam (>10^7 W/cm^2 peak power, 20 kw, beam divergence = 2×10^3 rad)
Specific power consumption	1,000 W/mm^3/min
Wavelength for machining operation	0.4–0.6 μm
Maximum metal removal rate	5 mm^3/min
Parameters affecting the process	Beam power intensity, beam diameter, workpiece, and melting temperature
Material application	All material
Shape application	Drilling the hole up to 250 μm, when the workpiece thickness exceeds 0.25 mm, a taper of 0.005 mm/mm is noticed
Limitations	Large power consumption; cannot cut materials with high thermal conductivity or high reflectivity
Process efficiency	Low 0.3 to 0.5%
Width of cut	2.5 to 9 mm

Table 3.7: Comparison between laser beam machining (LBM) and electron beam machining (EBM).

Parameter	Laser beam machining	Electron beam machining
Working principle	Uses a high-energy laser beam to melt, vaporize, or sublimate material	Uses a high-velocity electron beam to heat and vaporize the material
Energy source	Laser (light amplification by stimulated emission of radiation)	Electron beams are generated by a cathode and accelerated by a high-voltage electric field
Medium of operation	Can operate in air, inert gas, or vacuum	Requires a high vacuum (~10^{-5} mm Hg) to prevent electron scattering
Power density	10^6–10^{12} W/cm^2	10^6–10^9 W/cm^2
Material removal mechanism	Melting and vaporization	Vaporization due to localized heating
Focusing mechanism	Lenses or mirrors	Electromagnetic lenses
Depth of cut	Shallow (up to 10 mm)	Deeper cuts (up to 15 mm)

Table 3.7 (continued)

Parameter	Laser beam machining	Electron beam machining
Spot size	10–200 µm	2–50 µm (more precise)
Heat-affected zone (HAZ)	Larger HAZ due to heat conduction in air	Smaller HAZ due to operation in a vacuum
Beam deflection control	Optical mirrors for steering	Magnetic coils for precise beam control
Workpiece material	Suitable for metals, ceramics, polymers, and composites	Best for metals and high-melting-point materials
Accuracy & precision	High precision, but slightly lower than EBM	Extremely high precision is achieved due to the smaller beam diameter
Machining speed	Faster, due to continuous beam operation	Slower due to the need for pulsed electron beams
Applications	Micro-drilling, cutting, engraving, welding, and medical applications	Fine drilling, micro-machining, semiconductor fabrication, and aerospace components
Limitations	– Lower efficiency (~0.3–0.5%) – Large tapered holes in thick materials – HAZ is higher	– Requires a vacuum chamber (unsuitable for large workpieces) – Expensive and complex setup

Ultrasonic machining (USM) is a non-traditional machining process that removes material using high-frequency mechanical vibrations of an abrasive slurry against the workpiece. It is particularly useful for hard and brittle materials, such as ceramics, glass, and carbides.

– Working principle of USM
  – A tool oscillates at a high frequency (20–40 kHz) and an amplitude of 15–20 µm using ultrasonic energy.
  – Abrasive slurry (suspended particles such as silicon carbide or aluminum oxide) flows between the tool and the workpiece.
  – The abrasive particles impact the workpiece, causing microscopic chipping and gradual material removal.
  – Since USM is a mechanical erosion process, no heat generation occurs.
  – The material is removed by
    (i) the hammering of abrasive particles on the workpiece by the tool,
    (ii) the impact of free abrasive particles on the work surface,
    (iii) the erosion due to cavitations and chemicals.
– Components of USM: The main components of the ultrasonic machining system are shown in Figures 3.29 and 3.30. It consists of:

i.   Power supply: Converts electrical energy into high-frequency ultrasonic vibrations.
ii. Transducer: Converts electrical energy into mechanical vibrations (piezoelectric or magneto strictive).
iii. Tool holder/horn (acoustic head): Amplifies and transmits vibrations to the tool.
iv. Tool (sonotrode): Vibrates at an ultrasonic frequency and shapes the workpiece by impacting the abrasive slurry.
v.   Work table (fixture): Holds and secures the workpiece during machining.
vi. Feeding point: Directs the abrasive slurry toward the machining zone.
vii. Abrasive slurry and pump unit: Supplies the abrasive mixture (e.g., SiC, Al_2O_3) to the machining area for material removal.
viii. Workpiece: Typically hard and brittle materials, such as ceramics, glass, and carbides.

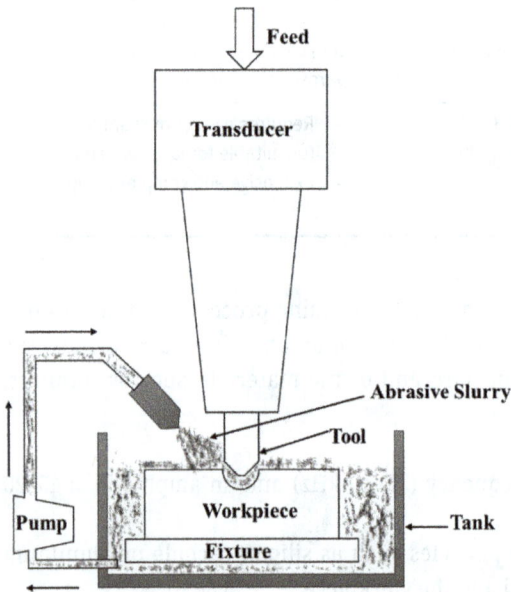

Figure 3.29: Ultrasonic machining setup.

- **Types of USM:**
  i. Rotary ultrasonic vibration machining
  ii. Chemical-assisted ultrasonic vibration machining
- **USM applications:**
  1. Since the cutting action occurs between the face of the tool and the workpiece, holes of any desired shape and dead-end cavities of any desired contour can be produced by shaping the tool face to the desired shape and contour.

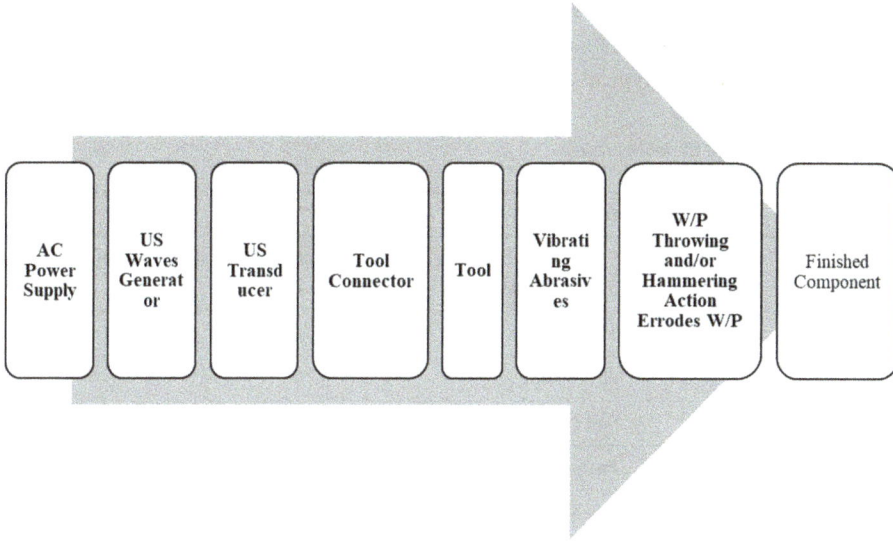

Figure 3.30: Flowchart showing the working process of USM equipment.

2. It can produce holes and cavities in materials of virtually any kind and hardness.
3. Holes from 0.025 mm to 125 mm in diameter and 37 mm deep are common.
4. Tolerance of 0.025 mm and surface finish of 15 µm can be obtained.
5. Cutting rate depends on the hardness and density of the workpiece, as well as the coarseness of the abrasive.
6. It can be used for very hard and brittle materials (as there are no thermal stresses).
7. It is quite frequently used for tool and die work.

– **Advantages of USM**
1. Hard and brittle materials can be machined into complex shapes with good accuracy and a reasonable finish.
2. Not affected by the electrical or mechanical characteristics of the work material.
3. Holes of any geometrical shape can be obtained.
4. No moving parts, no hazards.
5. Power consumption is about 0.1 W h/mm^2 for glass and about 5 W h/mm^3 for WC.

– **Limitations**
1. MRRS are slow
2. Depth of the hole produced is limited.
3. Tool wear is high, and sharp corners cannot be produced.

A summary of the USM process has been shown in Table 3.8, which provides highlighting key parameters such as the tool motion, abrasive medium, material removal mechanism, and typical applications.

Table 3.8: Summary of the USM process.

Parameter	Details
Metal removal mechanism	Brittle fracture due to the impact of abrasive, grains caused by tooling at ultrasonic frequency
Frequency range	15–30 kHz
Amplitude	25–100 µm
Tool material	Soft steel (stainless steel or low-carbon steel).
Abrasive used	Boron-carbide, silicon carbide, diamond (100–800 grit size)
Medium slurry	(30–40% concentration of abrasive)
Ratio of MRR/Tool wear rate	1.5 (for tungsten carbide workpiece), 100 (for glass workpiece)
Gap	25–40 µm
Critical parameters	Vibration frequency (20 kHz) and amplitude 15–20 µm, tool material, grit size 200–2,000. Abrasive type, slurry concentration, and slurry viscosity.
Application	
(a) Materials	Metal alloys, semi-conductors, non-metals (glass and ceramics)
(b) Shape	Round and irregular holes/impression
Limitation	Low MRRs, high tool wear rate, depth of holes, and small cavities

3.5.6 Abrasive jet machining (AJM)

In abrasive jet machining, a concentrated stream of abrasive particles, carried by high-pressure air or gas, is directed onto the work surface through a nozzle. The high-velocity abrasive particles erode the material, resulting in its removal.

The AJM apparatus can be used to determine the MRR for materials such as glass, ceramics, and aluminum sheets by varying parameters like pressure, NTD, and abrasive flow rate. In this project, we have designed and fabricated an abrasive jet machining apparatus for drilling holes and cutting glass, ceramic plates, and aluminum sheets. The structure and components of the AJM are represented in Figure 3.31.

– **Working principle**

AJM removes material by directing a high-velocity stream of abrasive-laden gas onto the workpiece. Micro-abrasive particles, propelled by an inert gas at speeds of up to 300 m/s, impact the surface, causing erosion, as shown in Figure 3.32. This process is widely used for cutting, etching, cleaning, deburring, polishing, and drilling.

Material removal occurs through a chipping action, making AJM particularly effective for hard and brittle materials such as glass, silicon, tungsten, and ceramics. However, soft and resilient materials like rubber and certain plastics resist this chipping effect and are not effectively processed using AJM.

Since AJM is a non-contact process, there is no workpiece chatter or vibration, allowing it to produce fine, intricate details on extremely brittle objects. The process generates minimal heat as the abrasive-laden gas carries it away, preventing thermal damage to the workpiece.

In AJM, fine abrasive particles (ranging from 0.015 mm to 0.06 mm in diameter) travel at velocities of 150–300 m/s under air discharge pressures of 2–10 atmospheres. The high-speed impact causes tiny brittle fractures, and the dislodged particles are carried away by the air jet. This makes AJM particularly suitable for machining brittle and fragile materials. The structured cutting mechanism is shown in Figure 3.33.

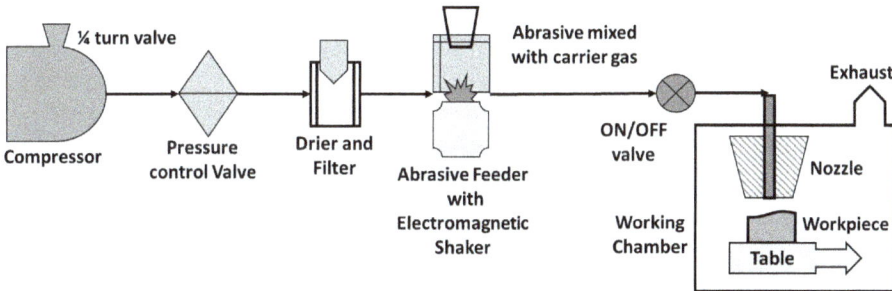

Figure 3.31: Schematic diagram of an abrasive jet machine with its components.

– **Setup of AJM**

A schematic layout of AJM is shown in Figure 3.31. The gas stream is then passed to the nozzle through a connecting hose. The velocity of the abrasive stream ejected through the nozzle is generally of the order of 330 m/s. The main components of AJM consist of:

i. Compressed air/gas supply: Provides the high-pressure carrier gas (air, nitrogen, or CO_2) to propel abrasive particles up to the pressure range of 2–10 bar (atmospheres).

ii. Pressure regulator: Controls and maintains the desired pressure of the carrier gas.

iii. Abrasive feeder: Stores and supplies fine abrasive particles (e.g., Al_2O_3, SiC, glass beads) into the air stream. It can be a gravity-fed or vibratory feeder system.

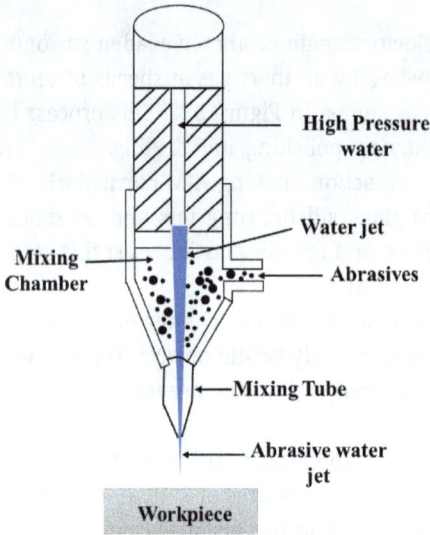

Figure 3.32: Schematic diagram of metal cutting in abrasive jet machining.

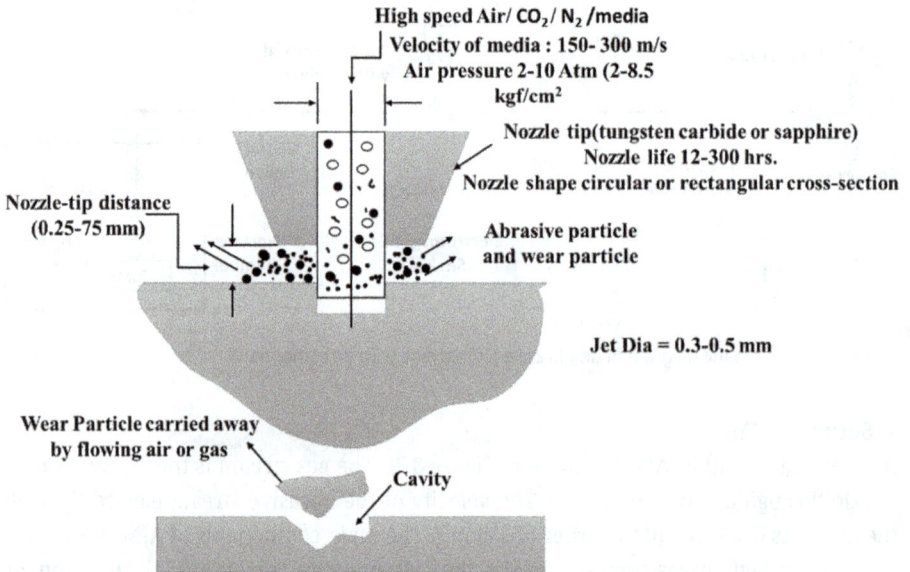

Figure 3.33: Mechanism of working in abrasive machining.

iv. Mixing chamber: Ensures uniform mixing of abrasive particles with the compressed air before being ejected through the nozzle.

v. Nozzle
   - Directs the high-velocity abrasive jet onto the workpiece.

- Made of wear-resistant materials such as tungsten carbide or sapphire.
- Typical nozzle diameter: 0.2–0.8 mm.
- Various types of nozzles in AJM are discussed in Figure 3.34. Whereas table 3.9 compares the diameter ranges and lifespan of round and rectangular-shaped nozzles made from tungsten carbide (WC) and sapphire, highlighting their respective performance characteristics.

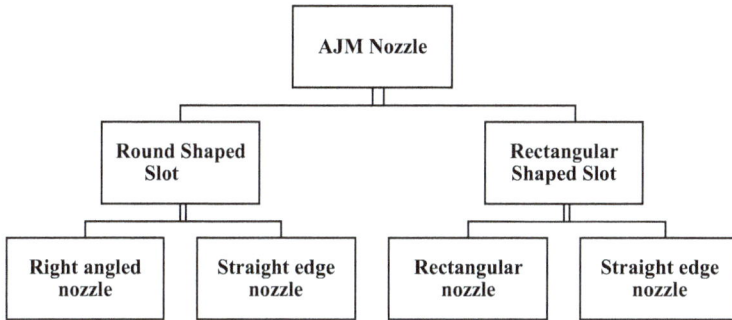

Figure 3.34: Various categories of abrasive jet machining nozzles.

Table 3.9: Comparison of Nozzle Materials and Their Specifications.

Nozzle material	Round shape nozzle dia (mm)	Rectangular shape nozzle dia (mm)	Life of nozzle (h)
Tungsten carbide (WC)	0.2–1.0	0.075 × 0.5 to 0.15 × 0.2.5	12 to 30
Sapphire	0.2–0.8	–	300

vi. Abrasives: Aluminum oxide (Al_2O_3), silicon carbide (SiC), glass beads, crushed glass, and sodium bicarbonate are some of the abrasives used in AJM. The selection of abrasives depends on MRR, the type of work material, and machining accuracy. Table 3.10 summarizes the types of abrasives used in Abrasive Jet Machining (AJM), highlighting their grain sizes and typical applications.

vii. Dust collector (filtration system): Removes airborne particles and used abrasives to maintain a clean working environment through the use of a vacuum or suction system.

viii. Workpiece holder/table.

ix. Exhaust/filter unit (optional): Further refines air quality by filtering out fine dust particles from the machining area.

Table 3.10: Types of abrasives used in AJM.

Abrasives	Grain sizes	Application
Aluminum oxide (Al$_2$O$_3$)	12, 20, 50 microns	Good for cleaning, cutting, and deburring
Silicon carbide (SiC)	25, 40 micron	Used for similar applications, but designed for hard materials
Glass beads	0.635 to 1.27 mm	Gives matte finish
Dolomite	200 mesh	Etching and polishing
Sodium bicarbonate	27 micron	Cleaning, deburring, and cutting of soft material. Light finishing below 500 °C

x. Additional elements (optional).
- Vibration mechanism: Helps control the abrasive flow in the feeder.
- Automation and CNC integration: For precise machining in advanced AJM setups.

- **Metal removal rate (MRR) in AJM:** MRR in AJM refers to the amount of material removed per unit time and is influenced by various parameters like abrasive particle size, velocity, pressure, and nozzle distance. MRR increases initially with an increase in the mixing ratio (mass fraction of abrasive) and then decreases as the particles collide with each other when dense. As the abrasive mass flow rate increases, MRR increases. Similarly, as the nozzle tip distance increases, MRR increases, i.e., the process efficiency improves up to a certain optimum distance; beyond this value, MRR remains unaffected by the increasing nozzle-tip distance. When this distance becomes excessive, the material removal rate starts decreasing.
  - **MRR formula for AJM**

$$MRR = C \cdot \rho \cdot V \cdot d^3$$

where C = Constant depending on material properties
ρ = Density of the workpiece material (g/cm^3)
V = Velocity of abrasive particles (m/s)
d = Diameter of abrasive particles (mm)
- Factors affecting MRR in AJM
  i. Abrasive particle velocity
     - Higher velocity increases the impact energy, enhancing erosion and MRR.
     - Velocity typically ranges between 150 and 300 m/s.
  ii. Abrasive particle size
     - Smaller particles provide finer machining but lower MRR.
     - Larger particles remove more material but reduce precision.

   iii. Abrasive flow rate
- Increasing the flow rate increases MRR but may cause excessive wear off the nozzle.
- Optimal flow rate needs to be maintained for efficiency.

   iv. Gas pressure
- Higher gas pressure increases particle speed and MRR.
- Typical pressure range: 2–10 bar.

   v. Nozzle tip distance (NTD)
- The distance between the nozzle and the workpiece affects the impact force.
- Optimal NTD is around 0.5–3 mm for effective material removal.

   vi. Workpiece material properties
- Brittle materials like glass, ceramics, and silicon have a higher MRR due to easy chipping.
- Ductile materials (e.g., aluminum) resist erosion, leading to a lower MRR.

- Typical MRR values
  - Glass and ceramics: 0.5–1.5 mm^3/min
  - Metals (aluminum, brass): 0.1–0.5 mm^3/min
- **Applications/benefits**

1. Applicable to brittle and fragile materials.
2. Metals machined: Hard and brittle metals; non-metals include germanium, silicon, glass, ceramics, and mica.
3. Especially suitable for machining thin sections.
4. Used for drilling, cutting, deburring, etching, and cleaning.

The key characteristics of the AJM process has been shown in Table 3.11, which includes the metal removal mechanism, operating medium, nozzle specifications, critical parameters, and suitable materials.

Table 3.11: Summary of abrasive jet machining characteristics.

Parameter	Details
Mechanics of metal removal	Brittle fracture, impingement of abrasive grain at a high level
Medium used	Air, CO_2, SiC
Abrasive normally used	Al_2O_3, SiC
Air velocity	150–30 m/s
Pressure	2–10 atmosphere

Table 3.11 (continued)

Parameter	Details
Nozzle (nozzle tip distance = 0.25–0.75 mm)	WC, sapphire, orifice area: 0.05–0.2 mm^2, life 12–300 h
Critical parameters	Abrasive flow rate and velocity, nozzle tip distance from the work surface, abrasive grain size, and jet inclination
Material application	Hard and brittle metals, alloys, and non-metallic materials, e.g., germanium, silicon, glass, ceramics, and mica
Shape (job) applications	Drilling, cutting, deburring, etching, cleaning
Limitation	Slow metal removal rate (40 mg/min or 15 mm^3/min)

3.5.7 Water jet machining

Water jet machining (WJM) is a non-traditional machining process that utilizes a high-pressure jet of pure water to cut soft and porous materials. It is commonly used in industries where avoiding thermal damage or mechanical stress on the workpiece is essential.

This process is primarily employed for cutting and slitting materials such as wood, foam, leather, and paper. Material removal occurs through the erosion effect of a high-velocity, small-diameter water jet, which efficiently cuts through the material without generating heat or causing structural damage.

– **Working principle of WJM:** WJM operates on the principle of high-pressure water erosion to cut materials. The process is purely mechanical, ensuring no thermal damage to the workpiece, which makes it suitable for delicate and soft materials.
  i. A high-pressure pump forces water through a small nozzle at extremely high velocities (600–1,000 m/s).
  ii. The high kinetic energy of the water jet impacts the workpiece, causing material erosion and removal.
  iii. Since there is no heat generation, the process is ideal for soft, porous, and delicate materials such as wood, foam, leather, and paper.

– **High-pressure water generation system:**
  – The oil pump is powered by an electric motor, which pumps oil into the system.
  – Water is then pumped into an intensifier, which uses low-pressure oil to generate high-pressure water.
  – The intensifier acts as a high-pressure pump, increasing water pressure up to 40 times that of the oil pressure.
  – This high-pressure water is directed through a small nozzle, creating a high-speed jet capable of cutting materials efficiently.

This method ensures precision cutting without introducing mechanical stress, making it a highly effective process in industries requiring clean, precise, and damage-free machining.

- **Components of the WJM system** (as shown in Figure 3.35)
    - i. High-pressure pump: Generates water pressure (2,000–4,000 bar).
    - ii. Water reservoir: Stores filtered water for machining.
    - iii. Nozzle: Converts high-pressure water into a high-speed jet (typically made of sapphire or diamond).
    - iv. CNC table: Holds and moves the workpiece precisely.
    - v. Drain and filtration system: Collects used water for recycling.

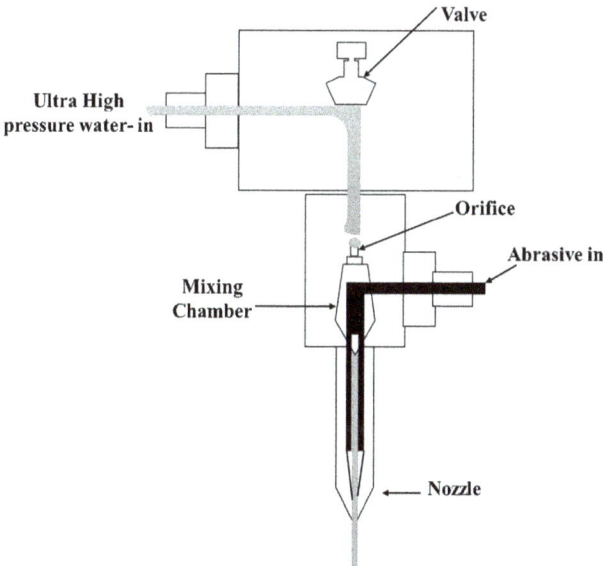

Figure 3.35: Schematic sketch of the water jet machining setup.

- **Calculating water pressure in WJM**

The water pressure (P_w) in WJM can be determined using the following relation:

$$P_w = (P_0 A_0)/A_w$$

where P_w = Water pressure after intensification (Pa or N/m^2)

P_0 = Initial oil pressure (Pa or N/m^2)

A_0 = Area of the oil piston in the intensifier (m^2)

A_w = Area of the water piston in the intensifier (m^2)

- **Explanation:**
    - The intensifier uses a hydraulic system in which low-pressure oil is used to generate high-pressure water.

- Since the oil piston has a larger area than the water piston, the pressure is amplified according to the area ratio.
    - The intensifier multiplies the pressure up to 40 times the oil pressure, producing the required high-pressure water jet.
- **Applications of WJM**
    - Food industry: Cutting cakes, vegetables, and frozen products.
    - Textile industry: Precision cutting of fabrics.
    - Paper and packaging: Cutting cartons and foam packaging.
    - Automotive and aerospace: Trimming soft composite materials.
- **Advantages**
    - Water is cheap, non-toxic, and can be easily disposed.
    - Water jet approaches the ideal single-point tool.
    - Any contour can be cut.
    - A clean, sharp cut can be obtained.
    - Does not generate heat; thermal degradation is not caused.
    - Best suited for explosive environments.
    - Dustless environment: Advantages for cutting asbestos, glass fiber insulation materials, which produce dust.
    - Noise is minimum during cutting
    - No moving parts; less maintenance.
    - Fluid can be reused by filtering.
- **Limitation**
    - Equipment is moderately expensive.
    - Not well-suited for hard, nonporous material.
    - Brittle materials cannot be machined.
    - Contaminated water must be treated before disposal.
    - High pressure usage requires safety considerations.

A comparative overview of AJM, WJM, and AWJM is presented in Table 3.12, highlighting their differences in working principles, mediums used, suitable materials, and common applications.

Table 3.12: Comparison between AJM, WJM, and AWJM.

Parameter	Abrasive jet machining (AJM)	Abrasive water jet machining (AWJM)	Water jet machining (WJM)
Medium used	High-velocity gas (air/N_2) with abrasive particles	High-pressure water with abrasives (SiC, Al_2O_3)	High-pressure pure water
Carrier fluid	Compressed air or gas	Water mixed with abrasives	Water only

Table 3.12 (continued)

Parameter	Abrasive jet machining (AJM)	Abrasive water jet machining (AWJM)	Water jet machining (WJM)
Abrasive particles	Al_2O_3, SiC, glass beads	Garnet, SiC, Al_2O_3	Not applicable
Material removal mechanism	Erosion by the impact of high-velocity abrasive particles	Erosion by high-velocity water and abrasives	Erosion by high-velocity water
Pressure range	2–10 bar	2000–6,000 bar	2000–4,000 bar
Velocity	150–300 m/s	500–900 m/s	600–1,000 m/s
Nozzle material	Tungsten carbide, sapphire.	Sapphire, diamond	Sapphire, diamond
Workpiece materials	Brittle materials (glass, ceramics, silicon, thin metals)	Hard metals, composites, thick materials	Soft materials (rubber, plastics, food, paper)
Heat generation	Low	Very low	None
Accuracy	Moderate (±0.1 mm)	High (±0.05 mm)	High (±0.1 mm)
Surface finish	Rough	Good	Very good
Applications	Cutting, etching, deburring, polishing	Cutting thick metals, aerospace, automotive	Cutting food, textiles, paper
Environmental Impact	Dust generation, requires dust collector	Uses water, needs filtration system	Eco-friendly but requires water disposal

- **Key differences:**
    - **AJM:** Best for **brittle materials** and small-scale machining, such as deburring, polishing, and drilling.
    - **AWJM:** More powerful than AJM, it is used for **thicker and harder materials**, such as metals and composites.
    - **WJM:** No abrasives, used for **soft materials** like rubber, textiles, and food cutting.

Questions

Long answer-type questions

1. Describe the working principle of a lathe machine. Discuss the different types of lathe operations and their applications.
2. Explain the construction and working of a CNC lathe machine. How does it differ from a conventional lathe?
3. Explain the working principle of a shaper machine. Describe the different types of shaper operations with suitable diagrams.
4. Discuss the various mechanisms used in a shaper machine for converting rotary motion into reciprocating motion.

5. Describe the construction and working of a planner machine. Compare the planner machine with the shaper machine in terms of design, operation, and application.

6. Explain the tool path and cutting speed in a planner machine. How does it impact the machining time and surface finish?

7. What are the different types of milling machines? Explain, with diagrams, the working principles of a horizontal milling machine and a vertical milling machine.

8. Discuss the various milling operations, such as plain milling, face milling, angular milling, and form milling. Provide examples of their industrial applications.

9. Explain the working principle and construction of a drilling machine. Describe the different types of drilling operations that can be performed using this machine.

10. Discuss the factors affecting drill bit life and the methods to enhance the performance and life of drill bits.

11. Describe the various types of grinding machines and their applications. Explain the working principle of a cylindrical grinding machine.

12. What are the different types of grinding wheels? Discuss the factors that influence the selection of a grinding wheel for a specific application.

13. Explain the working principle of electrical discharge machining (EDM). Discuss its advantages, limitations, and applications in modern manufacturing.

14. Discuss the various types of EDM processes, such as wire EDM and die-sinking EDM. How do they differ in terms of operation and applications?

15. Describe the working principle of electrochemical machining (ECM). Explain the advantages and disadvantages of ECM over conventional machining processes.

16. Discuss the various parameters that influence the material removal rate in ECM. How can these parameters be optimized?

17. Explain the working principle of electron beam machining (EBM). Discuss its applications in industries requiring high precision and accuracy.

18. Discuss the advantages and limitations of EBM. How does EBM compare with other non-traditional machining processes, such as laser beam machining (LBM)?

19. Describe the working principle of laser beam machining (LBM). Discuss the types of lasers used and their specific applications in material processing.

20. Explain the advantages and limitations of LBM. Discuss how it can be integrated into automated manufacturing systems for precision machining.

21. Explain the working principle of ultrasonic machining (USM). Discuss its applications in machining brittle and hard materials.

22. Discuss the factors affecting the material removal rate in USM. How can the process parameters be controlled to achieve optimal machining performance?

23. Describe the working principle of abrasive jet machining (AJM). Discuss the influence of abrasive type, size, and velocity on the material removal rate.

24. Explain the advantages and limitations of AJM in comparison to other non-traditional machining processes.

25. Compare and contrast the conventional machining processes, such as turning, milling, and grinding, with non-conventional processes, such as EDM, ECM, and LBM. Discuss their relative advantages and applications.

26. Discuss the environmental impact of different machining processes, including both conventional and non-conventional methods. What are the strategies to reduce the environmental footprint of these processes?

27. What are the emerging trends in machining processes? Discuss how advancements in automation, AI, and robotics are influencing the future of machining.

28. Discuss the importance of process selection in manufacturing. How do factors such as material properties, design requirements, and production volume influence the choice of machining process?
29. Explain the role of precision machining in the aerospace industry. Discuss how processes like EDM, LBM, and USM are used to meet the stringent requirements of aerospace components.
30. Discuss the application of non-traditional machining processes in the medical device industry. How do processes like ECM and EBM contribute to the manufacturing of complex medical implants and tools?

Short answer-type questions
1. What is the function of a tailstock in a lathe machine?
2. Explain the difference between turning and facing operations performed on a lathe.
3. How does the cutting speed affect the surface finish in a lathe operation?
4. What is the purpose of the quick-return mechanism in a shaper machine?
5. Explain the difference between a horizontal and a vertical shaper.
6. What is the role of the clapper box in a shaper machine?
7. How does a planer machine differ from a shaper machine in terms of size and application?
8. Describe the working principle of a double-housing planer machine.
9. What are the main advantages of using a planer over a shaper for large workpieces?
10. What is the difference between up milling and down milling?
11. Describe the process of indexing in a milling machine.
12. What are the applications of a vertical milling machine compared to those of a horizontal milling machine?
13. What is the purpose of using a center drill before performing a drilling operation?
14. How does a twist drill differ from a spade drill?
15. What factors influence the choice of cutting speed in drilling?
16. Explain the difference between surface grinding and cylindrical grinding.
17. What is the purpose of dressing and truing a grinding wheel?
18. How does the grade of a grinding wheel affect the material removal rate?
19. Compare the working principles of EDM and ECM.
20. What are the primary differences between EBM and LBM in terms of energy source and material removal mechanism?
21. Explain the role of abrasives in abrasive jet machining (AJM).
22. What is the principle of ultrasonic machining (USM), and which materials are best suited for this process?
23. Describe the significance of dielectric fluid in EDM.
24. What are the advantages of using ECM over conventional machining methods?
25. How does beam focus affect the machining process in laser beam machining (LBM)?
26. What are the typical applications of electrochemical machining (ECM) in the aerospace industry?

Objective questions
1. Which of the following machining processes is primarily used for producing external cylindrical shapes? **[GATE 2019]**
   (A) Milling   (B) Drilling   (C) Turning (Lathe)   (D) Planning
2. In a shaper machine, the cutting action occurs during the: **[SSC JE 2018, ME]**
   (A) Forward stroke
   (B) Return stroke
   (C) Both forward and return strokes
   (D) Neither forward nor return stroke
3. The tool used in a planer machine is typically: **[IES 2017, ME]**
   (A) Single-point cutting tool

    (B)   Multi-point cutting tool
    (C)   Abrasive tool
    (D)   None of the above

4.   Which of the following operations is commonly performed on a milling machine?

[GATE 2020, ME]

    (A) Drilling   (B) Knurling   (C) Slotting   (D) Boring

5.   Which of the following processes is most suitable for drilling deep holes with small diameters?

[IES 2018, ME]

    (A)   Conventional drilling   (B) Gun drilling   (C) Spot drilling   (D) Peck drilling

6.   Which machining process uses a grinding wheel as the cutting tool?   **[SSC JE 2019, ME]**

    (A) Turning   (B) Milling   (C) Drilling   (D) Grinding

7.   Which non-traditional machining process uses electric sparks to erode material from the workpiece?

[GATE 2018, ME]

    (A)   Electrochemical machining (ECM)
    (B)   Electrical discharge machining (EDM)
    (C)   Laser beam machining (LBM)
    (D)   Ultrasonic machining (USM)

8.   In Electrochemical machining (ECM), the tool material should be: **[IES 2019, ME]**
    (A)   Electrically conductive
    (B)   Electrically non-conductive
    (C)   Magnetic
    (D)   Non-magnetic

9.   Which process is used for precision drilling of very small holes using a focused electron beam?

[GATE 2021, ME]

    (A)   Laser beam machining (LBM)
    (B)   Electrical discharge machining (EDM)
    (C)   Electron beam machining (EBM)
    (D)   Ultrasonic machining (USM)

10. In ultrasonic machining (USM), the material removal rate is primarily dependent on: **[SSC JE 2017, ME]**
    (A)   Amplitude of vibration
    (B)   Frequency of vibration
    (C)   Abrasive material
    (D)   All of the above

11. In Abrasive jet machining (AJM), the material is removed by: **[IES 2020, ME]**
    (A)   Mechanical erosion
    (B)   Chemical reaction
    (C)   Thermal erosion
    (D)   Electrochemical reaction

Answers

1(C) 2(B) 3(A) 4(C) 5(B) 6(D) 7(B) 8(A) 9(C) 10(D) 11(A)

Chapter 4
Metal casting process and their applications

4.1 Introduction to foundry process

There are various methods by which we can shape metal into useful products. As we have already discussed in previous chapters, there are various types of operations like rolling, forging, drawing, extrusion, and machining used to manufacture products. Every operation has some limitations and requires specific characteristics of the material to process. For example, forging requires malleability, and drawing requires ductility. Brittle materials generally cannot be processed by the above methods. Therefore, to make useful products from brittle materials, the metal is first melted and poured into a desired-shaped cavity, where, after solidification, the metal assumes the shape of the mold cavity. Thus, the casting process consists of melting the metal, pouring it into the cavity, and allowing it to cool.

4.1.1 Casting terminology

The important terms used in metal casting with sand molds are briefly explained below (see Figure 4.1).
a) **Pouring basin:** A funnel-shaped cavity located at the top of the mold where molten metal is poured. It regulates the metal flow rate, minimizes turbulence, and prevents vortex formation at the sprue entrance.
b) **Sprue:** A vertical passage that directs molten metal from the pouring basin to the runner, ingates, and ultimately into the mold cavity. It ensures smooth and controlled metal flow without turbulence.
c) **Gate:** A channel through which molten metal enters the mold cavity.
d) **Riser:** Also known as a feeder head in foundry operations, it supplies additional metal to the casting to compensate for shrinkage during solidification.
e) **Runner:** A passage that carries molten metal from the sprue to the mold cavity through the gate.
f) **Pattern:** A replica of the final object to be cast, used to create the mold cavity.
g) **Parting line:** The dividing line between the two halves of the mold, strategically placed to ensure that the larger casting volume is in the drag section.
h) **Molding sand:** In foundry work, sand is prepared by mixing silica, clay, moisture, iron oxide, and aluminum oxide.
i) **Core:** A separate mold component made of sand mixed with organic binders and typically baked for strength. Cores are used to form holes and complex cavities within the casting.

https://doi.org/10.1515/9783112205891-004

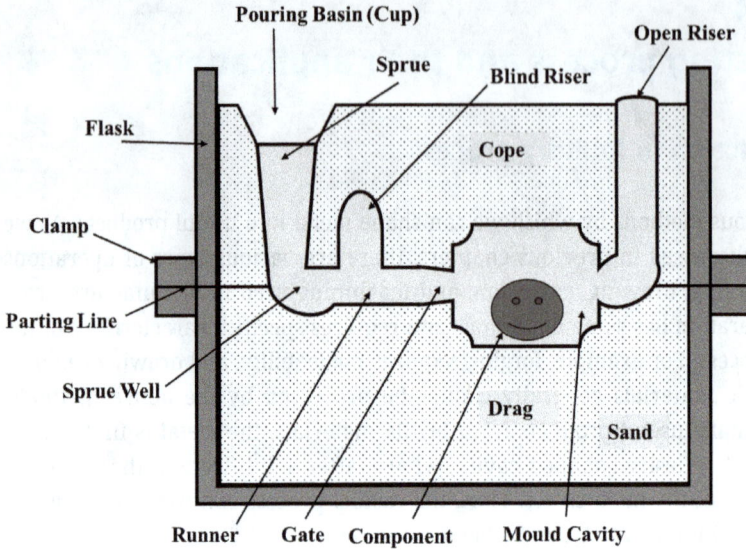

Figure 4.1: Cross-sectional view of the sand mold.

4.1.2 Steps in making sand castings

Figure 4.2 summarizes the basic steps involved in casting, which are explained below:

a) **Pattern making:** Create patterns using wood, metal, or plastic. It is a replica of the final product, so dimensional accuracy of the pattern is very important. Proper allowances must be given to the pattern.

b) **Prepare and test the necessary sand mixture:** The next step is to prepare a sand mixture. The sand mixture consists of silica, clay, moisture, additives, and iron oxides. These constituents must be mixed in the proper ratio. After preparation, test the sand mixture.

c) **Mold and core making:** Mold is nothing but a container made of sand that has a cavity in the shape of the object to be cast. A core is a body of sand (a projection in the cavity) that is employed to produce a cavity in the casting.

d) **Melting of metal:** Melting of metals is carried out in various types of furnaces, such as the cupola furnace.

e) **Pouring and solidification:** Pour the molten metal into mold cavity and allow it to solidify to obtain the final casting (product).

f) **Cleaning and finishing:** After removing the final casting from the mold, clean it.

g) **Inspection:** Inspect the casting and identify any defects in the casting, if present.

h) **Remove the defects, if any:** Remove the defects found during inspection using various methods that are available at the time.

i) **Heat treatment of cast product:** Heat treatment of the casting is performed to relieve internal stresses, refine the grain, and improve the mechanical properties.

j) **Re-inspect:** After heat treatment, re-inspect the casting.
k) **Packing:** And now the product is ready to sell in the market.

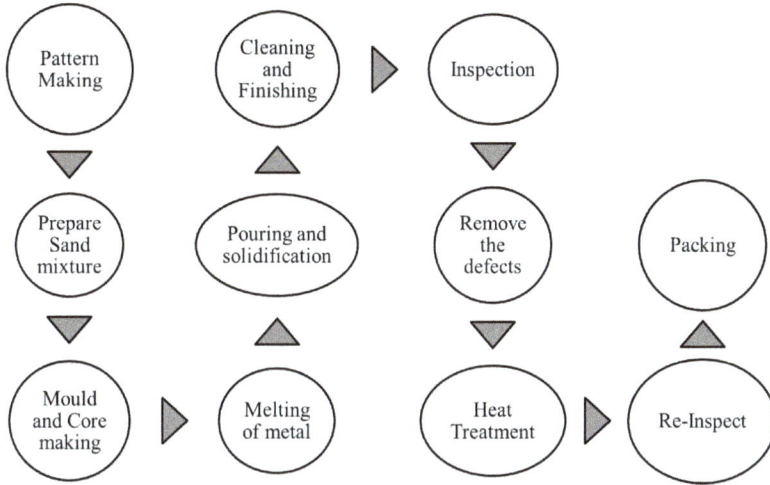

Figure 4.2: Basic steps involved in the casting process.

4.2 Pattern making

For making patterns, the patterns are first designed; the material for the pattern is selected, and molding procedure is kept in mind during the pattern design to make the molding process easier.

4.2.1 Pattern materials

Before discussing pattern making and types of patterns, first, we shall discuss the materials used for making patterns. Pattern materials can be classified into two major categories, which has been shown in table 4.1.

Table 4.1: Major categories of pattern material.

(A) Permanent category	(B) Disposable category
Wooden pattern	Wax
Metallic pattern	Polystyrene
Plastic pattern	Styrofoam
Rubber pattern	

Mostly wooden patterns are used because of the following properties.
a. Availability
b. Light weight
c. Easy to work on wood
d. Cheap than other material

4.2.2 Types of patterns

Various types of patterns used in the foundry process, as shown in Figure 4.3, are discussed below:

a) **Single-piece pattern or solid pattern:** A single-piece pattern, also known as a solid pattern, is the simplest type of pattern used in foundry work. It consists of a single, solid piece that directly replicates the shape of the final casting. This pattern is typically used for simple shapes and large castings where parting lines are not required or can be easily accommodated. It is suitable for casting large and heavy components such as machine bases, pipes, and manhole covers. See figure 4.3(a)

    Features of single-piece pattern:
– Made from a single solid block.
– Generally used for simple-shaped and large-sized castings.
– Suitable for flat surface molding, in which one side of the pattern rests on the molding board.
– Easy to manufacture and handle, it is cost-effective.

b) **Split pattern or two-piece pattern:** Patterns of complicated-shaped castings can't be made in a single piece due to some difficulties associated with molding operation. Therefore, the pattern is made in two pieces or split. It is also known as a cope and drag pattern, as shown in figure 4.3(b). Many patterns cannot be removed from the mold if made in a single piece. Dowel pins are used to maintain the alignment of the upper and lower parts.

c) **Gated pattern:** In this type of pattern, runners, gates, and risers are made an integral part of it. This type of pattern is used for the large-scale production of castings. The gated pattern is made of metal, which eliminates warpage or distortion during use. Many small castings are attached to a single sprue through gates and runners, as shown in Figure 4.3(c).

d) **Loose piece pattern:** A loose piece pattern as illustrated in figure 4.3(d), is specifically designed for castings with irregular shapes, where removing the pattern from the mold would otherwise be difficult. In such cases, the irregular portions of the pattern are made detachable using dowel pins or wires. This design allows the main pattern, which has a more regular shape, to be withdrawn first. The loose pieces can then be carefully removed through the gap left by the main pattern, ensuring that the mold remains intact without damaging the casting's contours. This type of pattern is particularly useful for castings with undercuts or projections, as it simplifies the molding process and prevents defects. These patterns are used for complex castings with projections such as gear housings, engine blocks, and pipe fittings, and they are ideal for patterns with internal cavities or intricate shapes that cannot be molded using a simple pattern.

Features of loose piece pattern:
- Consists of a main body with one or more detachable parts (loose pieces).
- The loose pieces are placed in the mold and held in position while the pattern is removed.
- Helps in creating intricate designs with undercuts or complex shapes.
- Requires careful placement and removal to avoid damaging the mold.

e) **Match plate:** This type of pattern helps to eliminate the mismatch between the cope and drag. It is generally used in machine molding. It consists of a match plate shown in figure 4.3(e), with half of the split pattern fastened on both sides of the plate. Accuracy in casting is very high with this type of pattern.

f) **Follow board pattern:** A follow board pattern illustrated in figure 4.3(f), is used for casting thin, fragile, or irregularly shaped components that are difficult to mold using conventional methods. It consists of a flat wooden board that supports the pattern during the molding process.

Since thin patterns are prone to collapsing during sand ramming, the follow board provides stability during molding:
- The thin pattern is placed on the follow board while the drag is packed with sand.
- The follow board is then carefully removed once the drag is packed.
- The rammed drag is inverted, the cope is placed on top, and the molding process continues.

This method ensures the pattern remains intact and helps create an accurate mold without deformation.

g) **Sweep pattern:** A sweep pattern is a form made on a wooden board that sweeps the shape of the casting in sand. It sweeps the sand by rotating about a post in mold. A regular cross-section is created by the sweep pattern. see figure 4.3(g).

h) **Segmental pattern:** This also generates a circular-shaped casting similar to the sweep pattern. However, it does not rotate continuously like the sweep pattern; instead, it creates mold in segments. See figure 4.3(h). First, it forms one part of the mold and then moves on to the next part, continuing this process till mold is completely prepared.

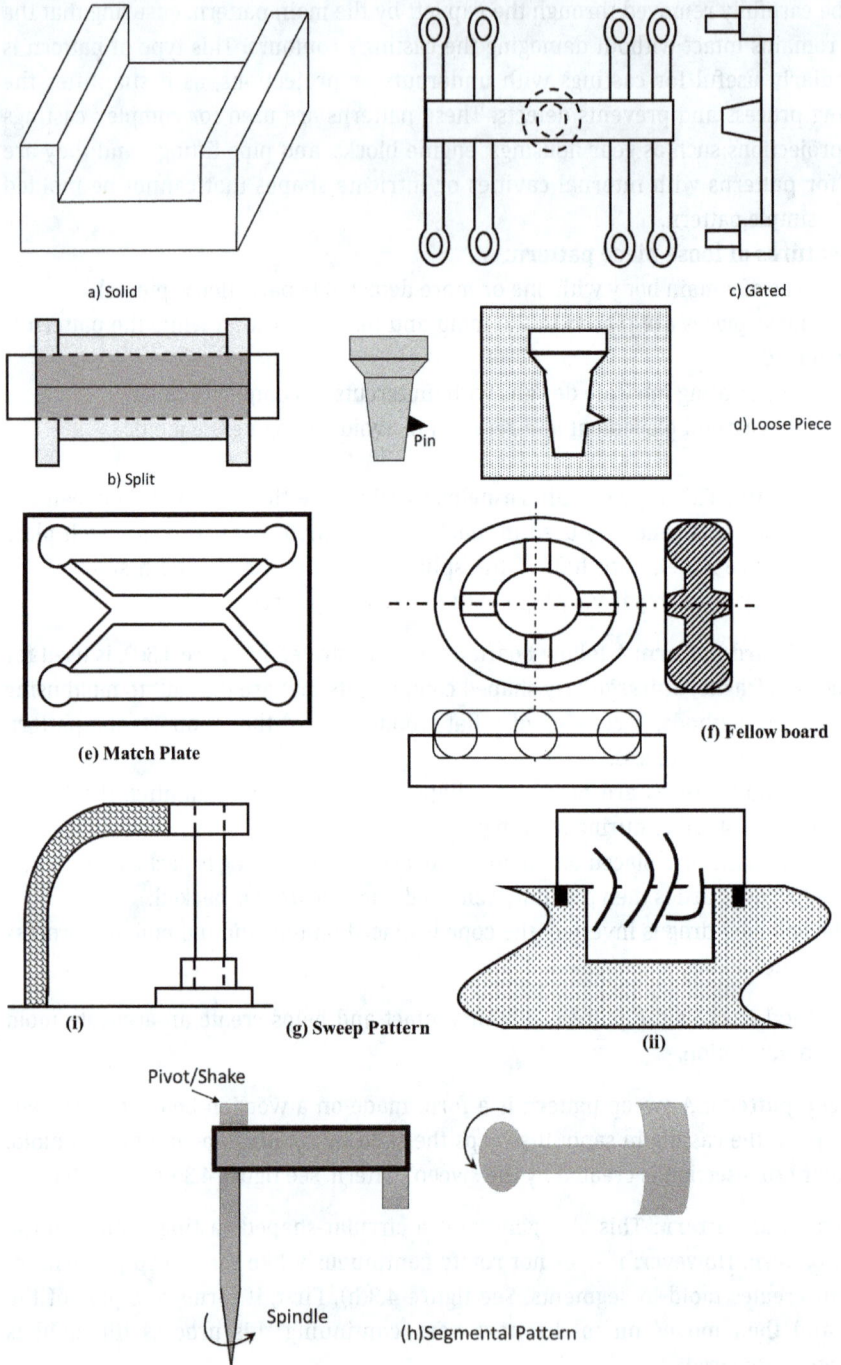

a) Solid

c) Gated

b) Split

Pin

d) Loose Piece

(e) Match Plate

(f) Fellow board

(i)

(g) Sweep Pattern

(ii)

Pivot/Shake

Spindle

(h)Segmental Pattern

Figure 4.3: Types of patterns.

4.2.3 Pattern allowances

Dimensions of the pattern differ from the final casting. This difference in dimensions is called allowances and is provided to ensure the easy withdrawal of the pattern and a dimensionally correct finished casting. There are mainly five types of allowances provided on a pattern.

a) **Shrinkage allowance:** After the pouring of molten metal, it solidifies, which leads to contraction in the metal. The contraction of molten metal occurs in three stages.

 i. Molten metal contraction from pouring temperature to freezing temperature is called liquid contraction.

 ii. After that, the temperature of the molten metal goes down, leading to the solidification of the metal, which is called solidifying contraction. During this contraction phase, a phase change occurs in the metal.

 iii. At last, when the final casting is taken out at room temperature, the casting contracts again. This is called solid contraction, which occurs at room temperature.

During these contractions, two types of contraction (liquid contraction and solidifying contraction) are compensated by risers and gates. The final contraction (solid contraction) is compensated by shrinkage allowances. All three stages of contraction are shown in Figure 4.4.

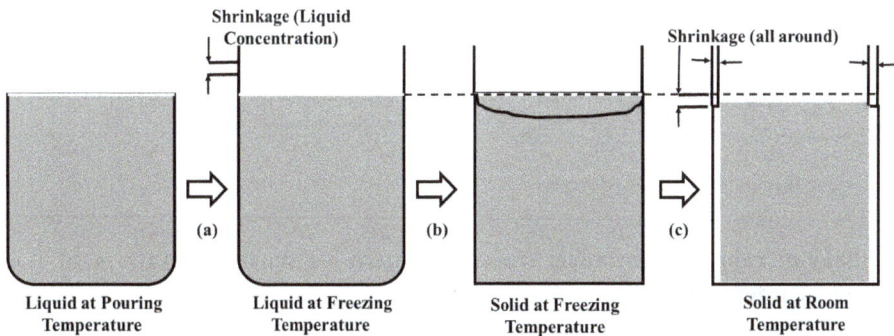

Figure 4.4: Stages of shrinkage as (a) liquid-liquid contraction, (b) liquid-solid contraction, (c) solid-solid contraction.

1. **Machining allowances:** After obtaining the final product from the mold, the surface finish of the product is not satisfactory. Therefore, to improve its surface finish, machining or finishing is required, which reduces the dimensions of the cast product. To compensate for this reduction in dimensions, machining or shrinkage allowance is provided on the pattern.

2. **Draft allowance**: It is the tapering provided on the sides of the pattern. Straight sides of the pattern cause cracks in mold cavity during the removal of the pattern from the mold. A slight inclination is provided on the pattern sides, which helps to reduce the draw force. Machine and draft allowances are indicated in Figure 4.5.

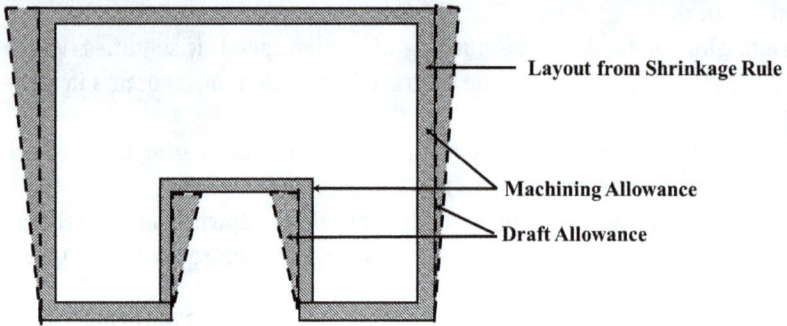

Layout from Shrinkage Rule

Machining Allowance

Draft Allowance

Figure 4.5: Draft and machining allowances.

3. **Distortion or camber allowance:** Distortion allowances are given to unsymmetrical shapes. When unsymmetrical shapes (U, T, etc.) solidify, they become distorted due to non-uniform solidification. To avoid this, the shapes of the pattern should be given an allowance of an equal amount in the opposite direction, as indicated in Figure 4.6.

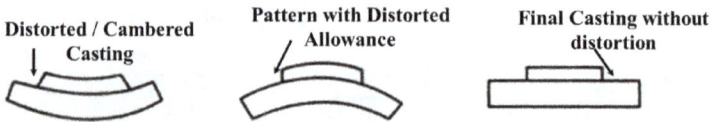

Distorted / Cambered Casting

Pattern with Distorted Allowance

Final Casting without distortion

Figure 4.6: Distortion or camber allowance.

4. **Shake or rapping allowance:** When the pattern is removed from the mold, it is shaken in order to make the surface free from mold wall. Due to this shaking, the size of mold cavity increases. To compensate for this, a small negative allowance is provided on the pattern. A schematic is shown in Figure 4.7.

**Shake or Rapping
Allowance**

Figure 4.7: Shake or rapping allowance.

4.3 Molding sand and its desirable properties

4.3.1 Constituents of molding sand

Commonly used sand compositions for ferrous metals and non-ferrous metals are given below in table 4.2:

Table 4.2: Major categories of molding sand and their properties.

Ferrous metals	Composition	Non-ferrous metals	Composition	
			Natural	Synthetic
Silica sand (AFS60–75)	83%	Silica sand (AFS60–75)	140	130
Clay	7%	Clay	15%	4–5%
Sea coal	6%	Moisture	8%	4%
Wood flour	0.7%	Permeability	15–20%	4%
Water (moisture)	3.3%	Green compression strength	0.4–0.55 kg/cm^2	0.55–0.7 kg/cm^2

4.3.2 Desirable properties of molding sands

Molding sand is an important element in the casting process because it forms the mold cavity used to shape metal components. Its quality has a considerable impact on casting accuracy, surface finish, and overall durability. The optimal molding sand should possess the following characteristics.

a) **Refractoriness:** It is the ability of sand to with stand high temperatures. High temperature means above the melting point of the final casting being produced. This property is found in silica and quartz sands. Rough grain size provides higher refractoriness.

b) **Green strength**: It is the strength of sand in a moist condition. This strength helps the mold retain its shape, prevents distortion, and prevents collapse.

c) **Dry strength:** The strength exhibited by molding sand in its dry state molding is called dry strength. When molding sand dries after molten metal is poured or loses moisture, it must possess dry strength to retain its shape, withstand the pressure of molten metal, and resist the erosive forces caused by the molten metal.

d) **Hot strength:** It is the strength of molding sand that allows it to resist maximum temperature. The ability of sand to retain its shape and prevent sand erosion at high temperatures is referred to as hot strength.

e) **Permeability/porosity:** After the pouring of molten metal into sand, the moisture, binder, and additives present in the sand produce steam and other gases. This property allows gases and steam to escape through the sand mold. If these gases do not easily escape, they may become entrapped and cause defects in the casting.

f) **Plasticity/flowability:** Plasticity is the property of sand by virtue of which it allows the sand to get compacted to a uniform density while ramming is done. By adding clay and water, flowability can be increased.

g) **Adhesiveness: i**t is the property of sand by which molding sand adheres to other surfaces, e.g., mold box.

h) **Cohesiveness:** It is the property of sand by which its individual grains remain bonded together and possess sufficient strength to withstand the pressure of metal flow.

i) **Collapsibility:** It is the ability of the mold or core to collapse (disintegrate) easily or automatically after the solidification of the casting, allowing free contraction of the metal, especially around the core, to protect the casting from tears and cracks caused by hindered contraction.

j) **Other requirements**: The molding sand should possess the following properties and additional characteristics have been summarize in table 4.3.

Table 4.3: Properties and additional characteristics of molding sands.

Ease of availability	Sand mixtures should have the following characteristics:	
Low coefficient of expansion	Green compression strength	1.2–1.3 kg/cm^2
Should not chemically react with the molten metal	Green shear strength	0.35–0.40 kg/cm^2
Should not stick to the casting surface	Moisture content	3.5–4.0%
Reusability and low cost	Permeability	80%
Thermal stability	Volatile combustibles	5–6%
Ease of sand preparation and control		

1. **Types of sand molds:** The major types of sand molds are outlined below:

a) **Green sand molds:** Green sand is sand mixed with moisture and clay. In order to prevent defects like voids in the cast product, proper moisture and permeability must be present in the sand, which only green sand can provide. It contains around 5% water and 16–30% clay. Green sand is used for small, medium, and simple-sized castings.

b) **Skin-dried sand molds:** They are prepared from clay-bonded sand, packed around the pattern to a depth of about 12 mm (skin depth). On drying, it forms a hard skin. Skin-dried molds can also be prepared by making a green sand mold and spraying a coating of binder, such as molasses, linseed oil, or gelatinized starch, which hardens with the application of heat. This process improves the surface strength and finish of the casting.

c) **Dry-sand molds:** They are made using coarse sand and binders and are completely dried. Surface finish improves, but due to the increased strength of the mold, the chances of hot tearing in hot-short metals increase. It requires more energy, space, time, and equipment.

d) **Loam molds:** These are used for large castings and are built from bricks. Mold face is plastered with thick loam mortar; sweep or skeleton patterns are used to give the final shape, and the mold is dried thoroughly, which takes time. These molds are not commonly used.

e) **CO_2 molding:** It uses washed, medium-sized silica sand mixed with 3–5% silica gel (water glass **$Na_2.xSiO_2 + nH_2O$**), a liquid. Upon completion of molding CO_2 gas is passed through the sand pores to form Na_2CO_3 and a gel of $xSiO_2.nH_2O$ based on the reaction:

$$Na_2O.xSiO_2 + H_2O + CO_2 \rightarrow Na_2CO_3 + xSiO_2.nH_2O$$

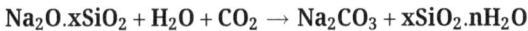

Silica gel

This gives a mold with strong wall faces (minimum wall movement). It, therefore, permits large precision castings to be produced.

f) **Cement bonded:** Its molds could be made using 10–15% cement for large castings in a pit. Cement forms a hard gel with sand, which makes it difficult for castings to be removed.

g) **Oil sands:** They consist of sand, a drying-type vegetable oil (such as linseed oil), and some cereal flour. This oil forms a polymer when heated to 230 °C, and the sand develops a bond.

h) **Resin-bonded sands:** Contain thermosetting resin binders (e.g., phenol formaldehyde, epoxy resin). Epoxy resins set at room temperature. The resin-coated sand is blown into the core box, heated to 200–250 °C, and the resin sets (hot box technique). It may also be cured with an airborne catalyst (cold-box technique). Originally developed for core making, it can also be used for molding. Phosphoric acid is used as a catalyst for the cold-box process.

2. **Types of molding:** Molding methods commonly used in modern industries are as follows:

a) **Bench molding:** Bench molding is carried out on a bench of convenient height. This method uses a small mold box and common tools used in molding. Bench molding is done for small-scale work.

b) **Floor molding:** The molding is carried out on the shop floor for medium and large work sizes that are not convenient to handle on a bench.

c) **Pit molding:** This is used for large-sized castings. The pit in the ground acts as the drag part of the molding flasks. A separate cope is used above it, containing the sprue and riser. The sides of the pit are brick-lined, and the bottom has a thick layer of cinders with connecting vent pipes leading to the floor level.

d) **Machine molding**: when molds are needed at a very fast rate, machine molding is used. Here, the machine eliminates all manual work with very efficient speed and accuracy that cannot be achieved manually.

4.4 Mold making with the use of a core

4.4.1 Element of mold

Elements of a mold are depicted in Figure 4.1, showing most of the elements in place. Molding using expendable (sand) molds and permanent patterns is briefly outlined in this chapter.

Sand molds are also called expendable molds as they are broken down to remove the casting. These molds are made from sand, while permanent molds are made from metals and will be discussed later in this chapter under the heading of die casting. Refractory material (sand mix) is packed around the pattern in the molding flasks. These flasks may be split into two pieces to accommodate the upper and lower halves of the patterns (see Figure 4.1). The upper flask is called the cope, and the lower flask is called the drag. In three-box molding for complex patterns, the central box is called the cheek. After reaming, the excess sand is leveled off with a straight bar, which is called a strike rod. Vent holes are made through the sand mold around the pattern to allow the escape of steam and gases generated when pouring molten metal into the mold cavity formed after the pattern is withdrawn.

4.4.2 Core making

When a casting needs to have a cavity or recess, some form of core is required to be introduced into the mold. A core is also defined as any projection of the sand into the mold cavity.

This projection may be formed during molding or made elsewhere and intro-duced into the mold after the pattern is withdrawn. Cores are also made of refractory materials like molds. They should have greater strength to permit safe handling but should collapse during solidification to allow for solid shrinkage of the hole for which the core has been used. This collapsibility prevents the casting from developing cracks due to solidification stresses and also permits easy removal of sand during the clean-ing of the holes.

Cores are molded in core boxes made of wood or metal. Cores may also be pre-pared in parts and then pasted together. Since cores are completely surrounded by molten metal, they must be properly vented to the outside to allow gases to escape.

1. **Types of cores:** The cores have been classified below and shown in figure 4.8.

a. Green sand cores	d. Drop core
b. Dry sand cores	e. Balanced sand cores
c. Vertical dry sand cores	f. Hanging sand cores

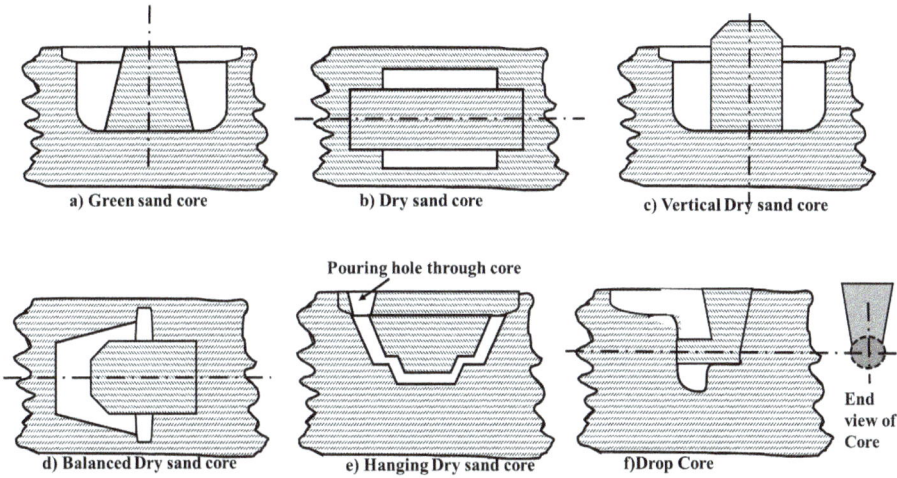

Figure 4.8: Typical cores: (a) green sand core, (b) dry sand core, (c) vertical dry sand core, (d) balanced dry sand core, (e) hanging dry sand core.

The green sand core should be used to keep the manufacturing cost low. Core boxes are made to prepare separate cores. A core is formed by ramming core sand (which con-sists of silica sand mixed with organic binders) into a core box or by using sweeps. Frag-ile or medium-sized cores are reinforced with wires to provide additional strength to withstand floating action or deflection during liquid metal pouring. In large cores, per-forated pipes are used; they provide strength and also serve as large vents. Round cores are generally made in two halves, which are then glued together after baking.

2. Core sand binders

The binders for core sand should be such that they provide ample strength during the pouring of liquid metal into the cavity and should burn or lose strength during the solidification of the casting. The core must disintegrate to allow for metal shrinkage around the casting cavity, thus protecting the casting from cracking around the cavities or holes thus made. To impart this characteristic, core binders are as follows:

a. Oils (such as linseed oil) as binders: A typical mixture consists of 40 parts sand and 1 part linseed oil, which forms a thin film around the sand grains that hardens when oxidized and baked for 2 h at 180–220 °C.
b. Binders soluble in water: e.g., wheat flour, dextrin, gelatinized starch, etc. The sand-to-binder ratio is 8:1.
c. Thermosetting plastics (phenol or urea-formaldehyde).
d. Furfural alcohol resin binders (do not require baking).
e. Sodium silicate binders (hardened by passing CO_2 gas through the core).

4.4.3 Molding procedure steps

Molding procedure and the steps involved are depicted in Figure 4.9. As a first step, half of a split pattern is placed on the molding board, sand is packed over it, and it is rammed in the drag half of the molding box. This drag is turned over in step 2, and the second half of the pattern is assembled over the first half. Sand is then filled in the cope with the sprue pin in place and rammed. Molding boxes are opened, and the gate is made. The core is placed in position, and the boxes are re-assembled. This is step 3, and the mold is now ready for pouring.

4.4.4 Principles of machine molding operation

To achieve optimum sand compaction, foundry molding employs a variety of ramming techniques, including jolt ramming, squeezing, and sand flinging, as shown in Figure 4.10.

a) **Jolt ramming:** In jolt ramming, the mold is lifted to a height of 50 mm and dropped back 50–100 times at a frequency of about 20 times per minute. This results in comparatively higher density at the bottom than at the top. For horizontal surfaces, it is satisfactory.
b) **Squeezing:** Gives more density in the top region than in the bottom region.
c) **Sand slinging operation:** It is very fast and results in uniform ramming density. This method requires a high initial cost.

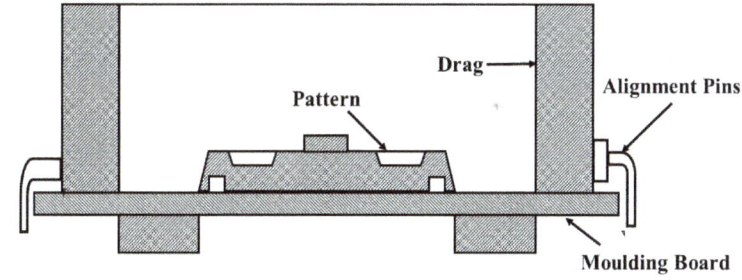

Step 1: Pattern placed on moulding board, sand filling and rammed in drag

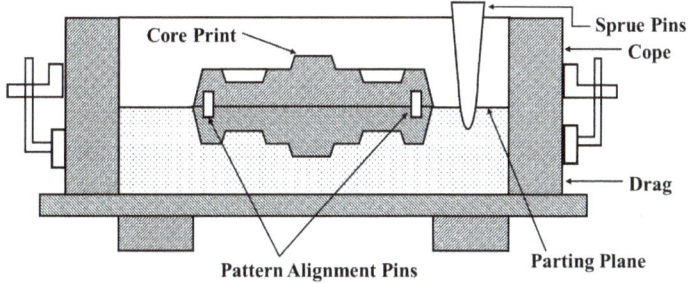

Step 2: Drag turned over and pattern assembled to ram cope with sprue pin in place

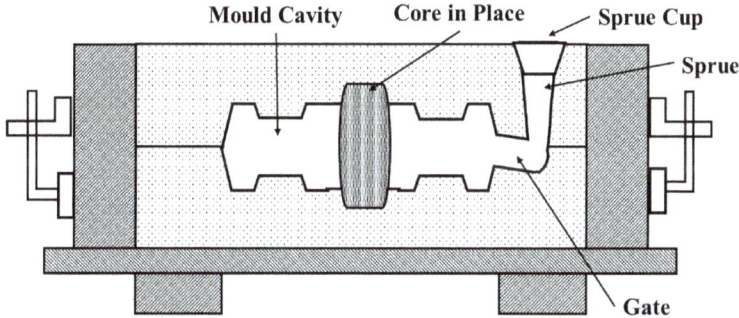

Step 3:Mould ready with dry sand core

Figure 4.9: Molding procedure (steps).

4.4.5 Classification of molds

The following is the classification of molds based on the materials used:

a) **Green sand molds:** Molds are formed by using damp molding sand.

b) **Skin-dried molds:** Two methods are used. In one method, sand around the pattern to a depth of about 10 mm is mixed with a binder. When it is dried, it will leave a hard, smooth surface on the mold. The rest of the mold consists of ordinary green sand. The other method is to prepare the entire mold with green sand

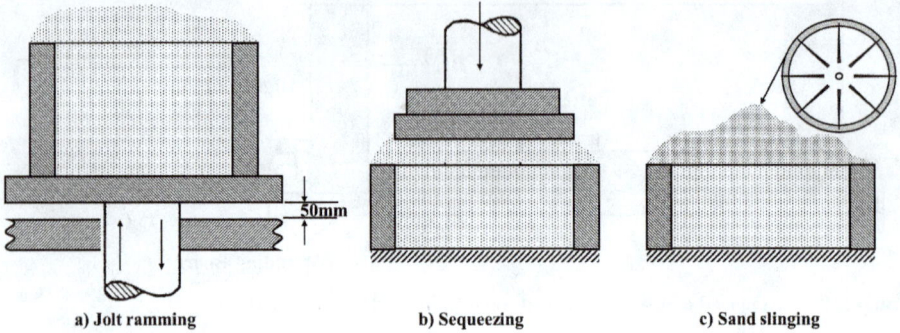

50mm

| a) Jolt ramming | b) Sequeezing | c) Sand slinging |

Figure 4.10: Machine molding types.

and coat its outer surface with a spray (or wash) that hardens when heat is applied. Sprays used include linseed oil, molasses, gelatinized starch, etc. The mold is then dried in both cases, either by air or by torch, to drive out excess moisture and harden the surface.

c) **Dry-sand molds:** These are prepared from fairly coarse molding sand mixed with linseed oil, as mentioned above, and are over-baked before use. These molds hold their shape when poured, and the castings are free from gas troubles due to moisture. Skin-dried and dry-sand molds are extensively used in steel foundries.

d) **Loam molds:** These are used for large castings. Initially, the mold is built using bricks and their pressure with a thick loam mortar. Mold shape is obtained with a sweep or skeleton pattern. Mold is then thoroughly dried to give it strength to withstand the heavy flow of molten metal. These molds take a long time thoroughly dried to give strength to stand heavy flow of metal. These molds take long time to prepare and are not commonly used.

e) **Furan molds:** These are used with disposable patterns and cores. Dry sharp sand is milled with phosphoric acid (accelerator). Furan resin is added, and mulling is continued to properly distribute the resin. The sand mix starts hardening but allows sufficient time to complete the mold.

f) **CO_2 molds:** These use clean sand mixed with sodium silicate as a binder. When CO_2 gas is passed through the mold it hardens. Very smooth, intricate castings can be obtained. The process, however, was originally developed for making cores.

g) **Metal molds:** These are mainly used in die-casting of low-melting-temperature alloys. Castings have good dimensional accuracy and a smooth surface finish.

h) **Special molds:** Plastics, cement, plaster of Paris, paper, wood, and rubber are all mold materials. These are discussed in special casting processes. With the above discussion of molds let us now look into the various methods of molding.

4.4.6 Methods of making molds

Mold-making techniques vary according to the size and quantity of castings needed. Common techniques are given below:

a) **Bench molding** (for small work) is done on a bench of convenient height.
b) **Floor molding** (for medium and large-sized castings) molding is done on the floor of the foundry due to handling problems.
c) **Pit molding** (for extremely large-sized castings) uses the pit as the drag of molding flasks.
d) **Machine molding** is used where a large number of molds need to be prepared at a fast rate. Its efficiency, speed, and accuracy are superior to hand molding.

4.5 Die casting

Die casting differs from sand casting, as temporary molds are mold used in sand casting, but in the case of die casting, permanent mold are used. There is no need to make mold again and again after casting. Die casting is a method of producing a desired product by injecting liquid metal into a metallic die under high pressure. Generally, it is used for the production of complex shapes and fine surfaces. Die casting is done by:

(i) Hot chamber   (ii) Cold chamber processes

4.5.1 Hot chamber die casting

A cylinder containing a pump and plunger is filled with liquid metal through a hole. When the plunger moves down in the cylinder, the liquid metal is forced into the mold cavity through the goose neck and nozzle. The pressure range is up to 35 MPa. The liquid metal is held in the die under pressure until it solidifies. The die is cooled by circulating water or oil through various passage ways in the die block. For zinc castings, the pumping pressure is up to 40 MPa. Refer to Figure 4.11(a).

4.5.2 Cold chamber die casting

In cold chamber die casting, molten metal is introduced into the injection cylinder. Here, the injection cylinder is not heated (as in hot chamber die casting, where the cylinder is heated by a furnace); hence, it is called cold chamber die casting. A measured quantity of liquid metal is poured into the refractory cylinder. The plunger moves and pushes this liquid metal into mold cavity under pressure ranging from 20 to 70 MPa. Refer to Figure 4.11(b).

a) Hot Chamber die casting

b) Cold Chamber die casting

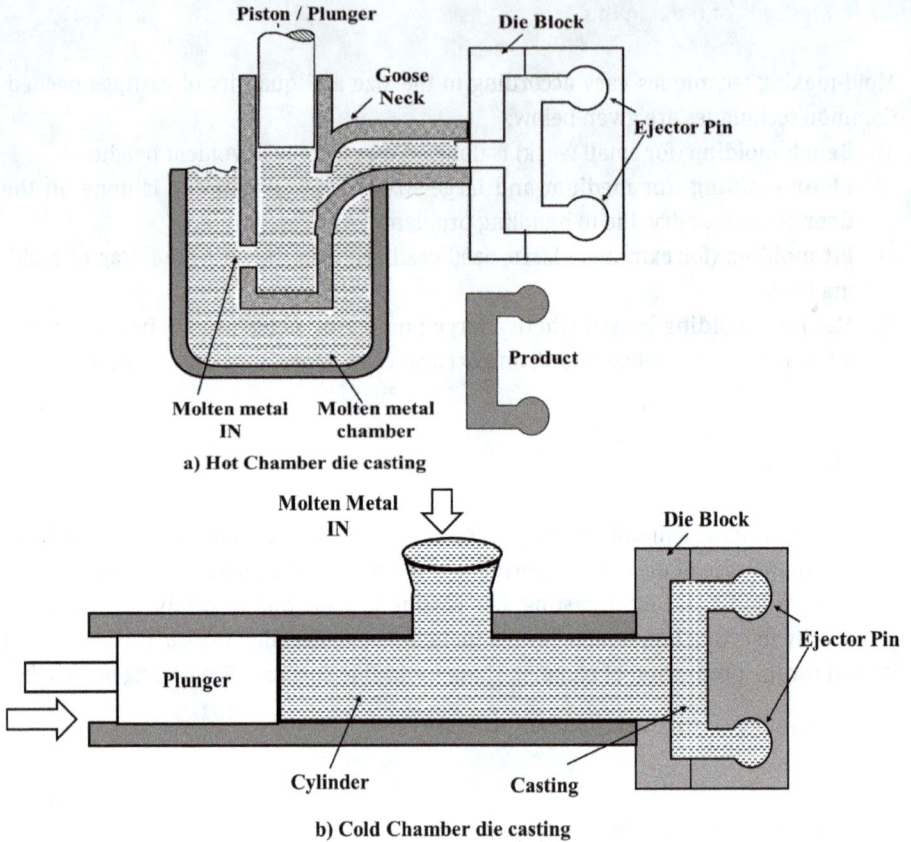

Figure 4.11: Types of die casting.

4.5.3 Centrifugal casting process

Centrifugal casting is a metal casting technique in which molten metal is poured into a rotating mold, allowing centrifugal force to distribute the metal and shape the casting. It utilizes inertial forces generated by rotation to distribute molten metal into the mold cavity. As the mold rotates at high speed around its central axis, centrifugal force pushes the molten metal outward, ensuring uniform filling and solidification. This process is ideal for producing hollow cylindrical components without the need for cores, such as pipes, bearings, and rings.

Types of centrifugal casting
1. **True centrifugal casting:** Used for symmetrical cylindrical components (e.g., pipes, tubes, and bushings).

Working: In true centrifugal casting, a metallic mold rotates continuously during solidification while molten metal is poured into it. The axis of rotation varies: it is horizontal for larger workpieces and vertical for shorter ones. To create hollow parts, the axis of rotation is aligned with the center of the casting. The high rotational speed generates a centripetal acceleration of about 60 g to 75 g, ensuring uniform metal distribution and dense, defect-free castings.

When multiple mold cavities are arranged around a central axis to produce multiple castings simultaneously, the process is called centrifugal casting.

2. **Semi-centrifugal casting:** Used for components like wheels and pulleys, where external detail is required.
3. **Centrifuging (centrifugal die casting):** Used for smaller, non-symmetrical parts by placing multiple cavities around the mold.

Process Steps

– **Mold Rotation:** A **mold is rotated** at high speed around its central axis.
– **Metal Pouring:** Molten metal is **poured into the rotating mold** and spreads outward due to centrifugal force.
– **Solidification:** The metal cools and solidifies while still rotating, ensuring a **uniform grain structure** and **high strength** in the final product.
– **Mold Removal:** Once solidified, the casting is removed and undergoes finishing operations.

This step-by-step process of centrifugal casting is illustrated in Figure 4.12.

Advantages

– Produces **high-strength, defect-free** castings.
– Eliminates **impurities and gas porosity** by pushing them toward the inner surface.
– **No need for cores** in hollow cylindrical shapes.
– **Excellent grain structure** and **mechanical properties** are achieved due to centrifugal force.

Limitations

– Limited to **cylindrical and axisymmetric** shapes.
– **High mold rotation speed** is required.
– **Expensive setup** compared to traditional casting methods.

Application

– Centrifugal casting is widely used in industries such as **automotive, aerospace, and manufacturing** for producing **high-performance metal components**.

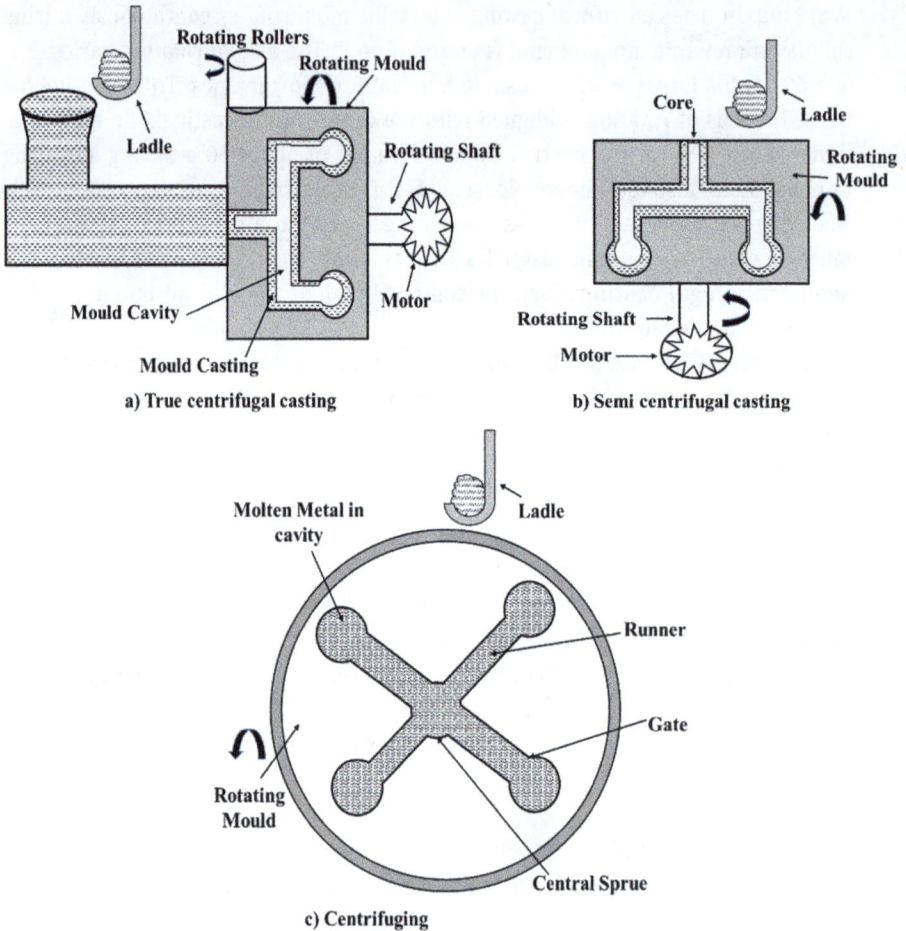

Figure 4.12: Centrifugal casting process variations shown as: (a) true centrifugal casting, (b) semi-centrifugal casting, and (c) centrifuging.

4.5.4 Investment casting

Investment casting, also known as the lost-wax process, is a precision casting method used to produce intricate and highly detailed metal components. This process is ideal for manufacturing complex parts such as gears, cams, and valves. It can also be used to cast large components, with diameters reaching up to 1.5 meters and weights of up to 1,140 kg.

The process involves the following steps:
1. **Pattern creation**: A wax model of the desired part is created, either through injection molding or manual shaping.

2. **Assembly**: Multiple wax patterns are attached to a central wax sprue to form a tree-like structure for batch casting.
3. **Ceramic shell formation**: The wax pattern is repeatedly dipped into a ceramic slurry and coated with fine refractory material to create a strong shell.
4. **Wax removal**: The ceramic-coated assembly is heated in an oven or autoclave at 700–1,000 °C to melt and remove the wax, leaving behind a hollow mold. This process ensures the mold can withstand the molten metal and improves fluidity during pouring.
5. **Metal pouring**: Molten metal is poured into the preheated ceramic mold, filling the cavity left by the wax.
6. **Shell removal**: After the metal solidifies, the ceramic shell is broken away to reveal the final casting.
7. **Finishing**: The casting undergoes cleaning, machining, and surface treatment, if necessary.

Such process steps, as depicted in figure 4.13, result in highly precise castings with excellent surface finishes, making investment casting ideal for industries such as aerospace, automotive, medical manufacturing, and jewelry due to its ability to create high-quality, intricate parts with tight tolerances.

Advantages of investment casting:
– **High precision and detail:** Capable of producing intricate designs and thin sections.
– **Excellent surface finish:** Minimal need for machining.
– **Wide material choice:** Can be used for ferrous and non-ferrous metals.
– **No parting lines:** Unlike sand casting, it produces seamless components.
– **Ideal for complex shapes:** Suitable for aerospace, medical, and jewelry applications.

Limitations of investment casting:
– **Expensive and time-consuming:** Requires detailed wax patterns and multiple steps.
– **Limited size:** Best for small to medium-sized components.
– **Fragile ceramic molds:** Can break if not handled properly.

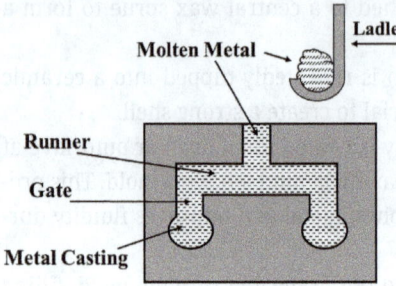

Figure 4.13: Investment casting process steps (also called the "lost wax" process).

4.6 Gating system

The gating system is a system that provides a passage for molten metal to enter into the mold cavity. After filling the mold cavity completely, any mold extra material rises in the riser. The basic structure of a gating system is represented in Figure 4.14.

The gating system consists of:

1. Pouring basin	4. sprue
2. Runner	5. Gate
3. Riser	6. Runner extension

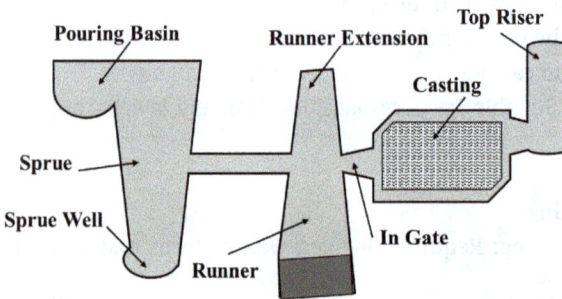

Figure 4.14: Components of the gating system.

4.6.1 Basic purpose of gating system

A successful gating system should fulfill the following requirements:
1. It should fill the mold cavity completely before the solidification of the molten metal.
2. It should avoid sudden changes in the direction of liquid metal movement. Due to sudden changes in the direction of molten metal, there will be (a) mold erosion, (b) turbulence, and (c) gas pick-up.

3. Help to promote directional solidification.
4. Non-metallic inclusions (slag, broken sand, etc.) should be separated by providing suitable traps.
5. Proper regulation of the metal entering the mold cavity should be ensured.
6. It should be practical and economical to make.
7. Should consume a minimum amount of metal.

4.6.2 Types of gating systems

a) **Top gating system:** In a top gating system, molten metal enters the mold cavity from the top. A proper temperature gradient is formed because the hottest metal remains at the top of the mold cavity. This temperature gradient facilitates directional solidification from the bottom toward the riser. In this type of gating system, there is no need for an additional riser, as the sprue itself acts as a riser.

Limitations:
- Entrapped oxides.
- Erosion of mold takes place.
- Due to turbulence in the gate, dross and gases occur.
- Not recommended for metals that oxidize easily.

b) **Bottom gating system:** In a bottom gating system, the entire liquid metal enters the mold cavity from the bottom of the mold cavity. This design reduces erosion (due to the reduction in turbulence) of mold and provides a smooth flow of liquid metal. Figure 4.15 depicts the different types of gating systems.

Limitations:
1. Favorable temperature is not achieved due to the gradual cooling of molten metal and when it rises.
2. Directional solidification is adversely affected because the hottest metal remains at the bottom while the coolest metal stays at the top.

c) **Parting line gating system:** These gates are a combination of the top gating system and the bottom gating system. In this system, metal enters from the centerline of mold (parting line). These gates are commonly used for all types of casting except for very deep mold.

Limitations:
1. Not used for deep mold, because in the case of deep mold, the metal has to fall onto the mold bottom through a considerable distance between the gate and the mold.
2. Causes erosion or washing of mold due to the metal entering from such a height.

3. In non-ferrous metals, this drop promotes dross formation and air absorption by the metal, leading to inferior casting quality.

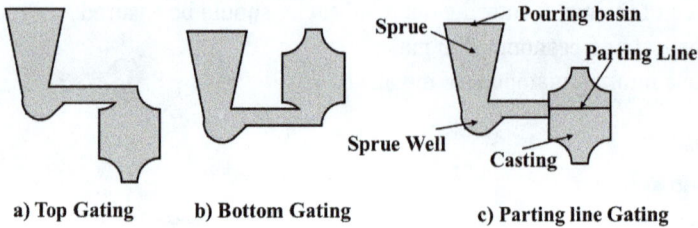

a) Top Gating   b) Bottom Gating   c) Parting line Gating

Figure 4.15: Types of gating system.

4.7 Cupola furnace

A cupola furnace is a vertical, cylindrical furnace used for melting metals, primarily for casting iron. It operates on the countercurrent heat exchange principle, where the charge materials – pig iron, scrap metal, limestone, and coke – are loaded from the top, while air is blown in from the bottom. The majority of gray and white cast iron is melted in a cupola furnace, which consists of a vertical cylindrical steel shell lined with refractory material. The furnace is supported by heavy steel or iron legs, elevating its lower end 3 to 5 feet above the ground. The bottom of the cupola is hinged to facilitate cleaning and is covered with a sand bed, sloping toward the tap hole (25–50 mm in diameter) at the pouring spout. Figure 4.16 shows a schematic diagram of the cupola furnace.

4.7.1 Construction and working of cupola furnace

– The furnace consists of a steel shell lined with fire-resistant refractory bricks to withstand high temperatures.
– The charge materials (metal, flux, and coke) are added in alternating layers.
– Air is blown into the furnace through tuyeres (air inlets) located near the bottom, aiding in combustion and increasing the furnace's temperature.
– The coke burns, generating intense heat that melts the metal.
– The molten metal collects at the bottom and is periodically removed through a tapping hole, while slag (impurities) is drained off separately.

4.7.2 Operation of a cupola furnace

i. **Initial setup**:
   - A bed of kindling wood is placed at the bottom, followed by a layer of coke (0.5–1.2 m deep).
   - The wood is ignited, and the coke bed is gradually built up.
   - Natural air draft enters through tuyeres positioned 0.4–0.5 m above the sand bottom.
ii. **Charging the furnace**:
   - Once the coke is fully ignited, alternating layers of iron (pig iron or scrap) and coke are added through the charging door, which is located midway up the furnace.
   - A typical charge consists of 1 part coke per 8–10 parts iron (by weight).
   - Limestone or flux materials are added to enhance fluidity and remove impurities.
   - Alloying materials may be introduced at this stage or in the ladle during tapping to modify the final composition.
iii. **Melting process**:
   - The charge is heated under natural draft for 30 min, then the blower is started to introduce forced air, raising the temperature.
   - The metal melts as it descends through the coke, with final melting occurring in the zone just above the tuyeres.
   - Molten iron begins to appear within 10 min after starting the air blast.
   - A clay plug temporarily seals the tap hole to contain the molten metal.
iv. **Tapping and slag removal**:
   - The molten metal collects at the base of the furnace, while the slag floats on top and is removed through a slag hole.
   - Once enough metal accumulates, the air blast is paused, and the tap hole is opened to drain the molten iron into holding or pouring ladles.
   - After pouring, the tap hole is resealed, and the air blast is resumed for continuous operation.
v. **Continuous operation**:
   - The cupola operates as a semi-continuous batch process, with iron and coke replenished regularly through the charging door.
   - The process can last up to 14–16 h.

4.7.3 Technical specifications

Flux materials such as limestone ($CaCO_3$), fluorspar (CaF_2), or soda ash (Na_2CO_3) help prevent oxidation and improve slag fluidity. Typically, 40 kg of limestone per ton of iron is used.

Air requirements: 5.78 m^3 of air at 100 kPa and 15 °C is required to melt 1 kg of iron.

Blast pressure: Typically ranges between 1,200 and 2,000 MPa.

Melting capacity: A continuously operating cupola can produce up to 120 tons of molten iron per hour.

Summary: The flux used is limestone (CaCO$_3$), fluorspar (CaF$_2$) or soda ash (Na$_2$CO$_3$) which protects iron from oxidation and makes slag more fluid. Approximately 40 kg of limestone is required per megagram (Mg) of iron. The air requirement is 5.78 m^3 of air at 100 kPa and 15 °C to melt 1 kg of iron. The blast pressure ranges from 1,200 to 2,000 MPa, and furnace capacities of 8–10 Mg/h/m^2 of furnace cross-section are available. In continuously run cupolas, it is not uncommon to produce 120 tons of molten iron per hour. This method remains one of the most efficient and widely used processes for melting iron in foundries due to its high productivity, cost-effectiveness, and continuous operation capability.

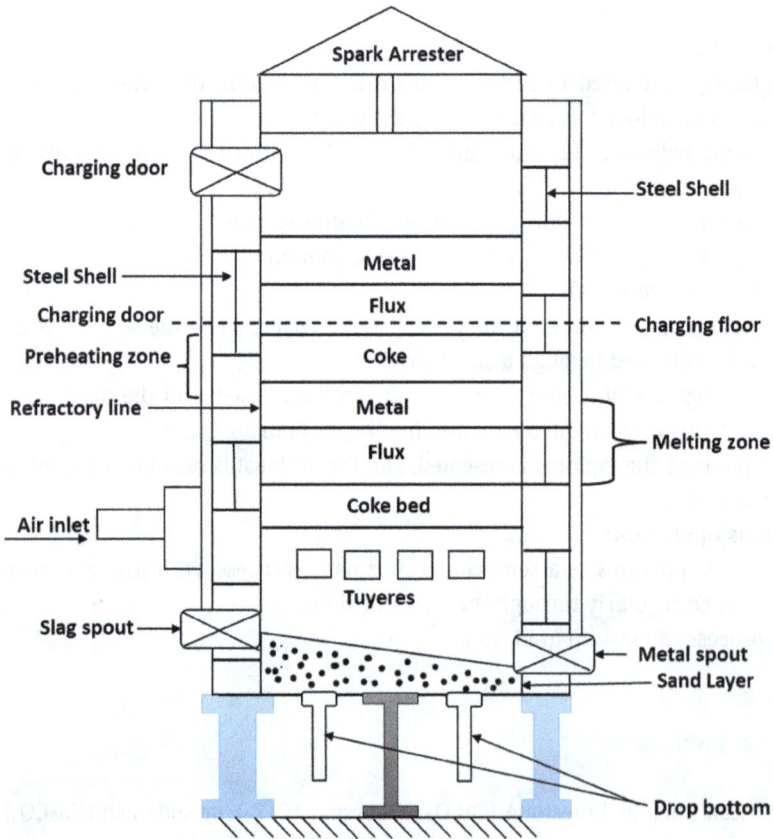

Figure 4.16: Cupola furnace.

– **Advantages of cupola furnace**
    Continuous operation: Allows for uninterrupted melting over long periods.
    High efficiency: Uses low-cost fuel (coke) while maintaining high melting rates.
    Simple construction: Easy to operate and maintain.
    Cost-effective: Ideal for large-scale iron casting industries.
    Controlled composition: The alloy content can be adjusted by adding different scrap metals.

– **Limitations of cupola furnace**
    Limited to ferrous metals: Primarily used for cast iron; not suitable for non-ferrous metals.
    Environmental concerns: Generates dust, CO_2, and other emissions.
    Requires skilled labor: Proper operation and monitoring are essential for efficient performance.

4.8 Casting defects

The defects are discussed in the order of their importance.

4.8.1 Surface defects

a) **Wash:** It is a small projection on the drag face of a casting. This defect occurs when molten metal is poured; due to improper flow of molten metal, sand is washed away. This washed sand appears at some other place as a sand inclusion, as shown in Figure 4.17(a).
b) **Rat tails or buckles:** When high-temperature molten metal is poured into mold cavity, the outer layer of the sand mold expands and separates from the sand behind it. This separation appears on the casting surface, where the casting surface shows a shallow indentation called a rat tail, as shown in Figure 4.17(b).
c) **Swell:** Swell occurs due to the enlargement of mold cavity caused by ferrostatic pressure, as shown in Figure 4.17(c).
d) **Misrun:** Due to the incomplete filling of mold caused by low pouring temperature, this defect occurs. Some additional reasons for these defects include improper liquid metal supply, incorrect gating design, etc. When the pouring temperature is too low or the casting section is too thin, there is a high chance of the molten metal starting to solidify before filling the mold cavity, which leads to a misrun, as shown in Figure 4.17(d).
e) **Hot tear:** These are ragged-edge cracks that occur due to tensile stress developing during solidification. It has been shown in Figure 4.17(e).

f) **Shrinkage:** Shrinkage in casting refers to the reduction in the volume of metal as it cools and solidifies, leading to potential defects such as voids or cracks in the final product. It is shown in Figure 4.17(f).

g) **Shift/mismatch:** It occurs due to the displacement of individual parts of the casting caused by incorrect assembly of the cope and drag of mold cavity with respect to each other, as shown in Figure 4.17(g).

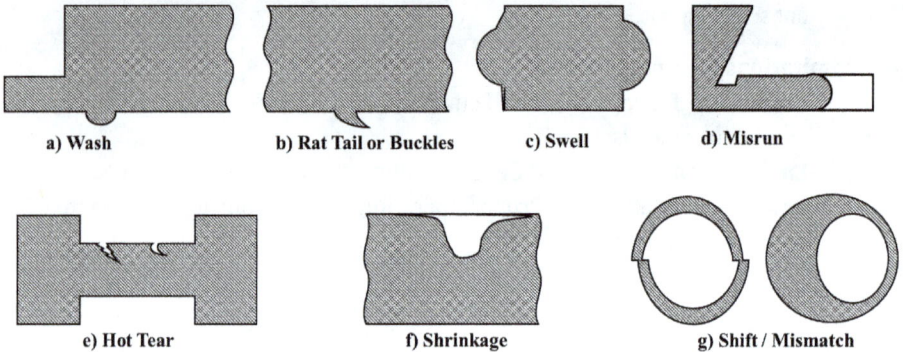

a) **Wash**   b) **Rat Tail or Buckles**   c) **Swell**   d) **Misrun**

e) **Hot Tear**   f) **Shrinkage**   g) **Shift / Mismatch**

Figure 4.17: Various casting defects appear on the surface, named as (a) wash, (b) rat tail, (c) swell, (d) misrun, (e) hot tear, (f) shrinkage, and (g) shift/mismatch.

4.8.2 Internal defects

a) **Blow holes:** These are simple round holes that occur on the surface of a casting and are termed open blows. Some blow holes occur below the surface of the casting and are also referred to as blow holes. When excess moisture is present in molding sand, it develops gas bubbles known as blowholes. See Figure 4.18(a).

b) **Porosity:** This defect is due to gases absorbed by the molten metal. Nitrogen, oxygen, and hydrogen are the main gases absorbed by molten metal during casting. It is shown in Figure 4.18(b). Porosity may also be caused by shrinkage, and it is called shrinkage porosity. Excessively high temperatures can lead to porosity.

c) **Cold shut:** A cold shut is an interface within a casting that forms when two metal streams meet without complete fusion. This defect occurs when the metal streams meet at temperatures too low to cause fusion, as shown in Figure 4.18(c).

d) **Pinholes:** These are small gas cavities formed due to trapped gases during solidification, as shown in Figure 4.18(d).

e) **Dross:** It refers to impurities or oxides that float on the surface of molten metal and can become trapped in the casting, causing defects, as shown in Figure 4.18(e).

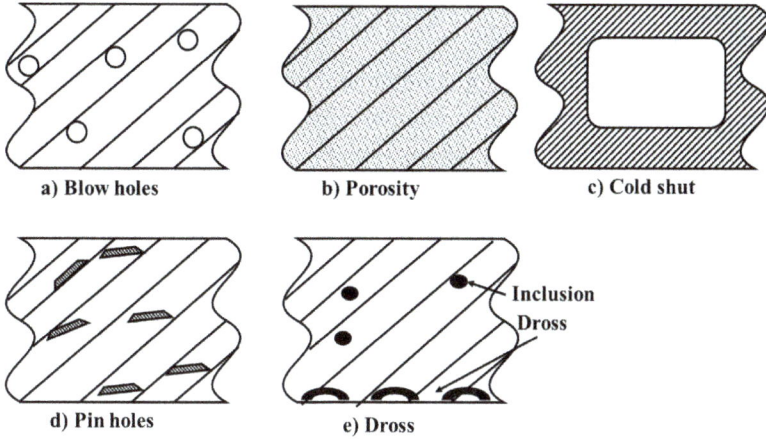

| a) Blow holes | b) Porosity | c) Cold shut |
| d) Pin holes | e) Dross |

Figure 4.18: Various casting defects (internal).

4.8.3 Applications of casting processes

Casting is one of the most commonly used and oldest known manufacturing processes. Some of the major applications are as follows:

a) Brass statues, idols, church bells, brass buttons, flower vases, gold crowns, etc.
b) Beds of machine tools, heavy-duty presses, flywheels, rolling mill stands, and cane crushers.
c) Electric motor castings, pumps, turbines, compressors, and hand pumps.
d) Gear boxes, heavy-duty gears, planer tables, IC engine castings, carburetors.
e) Manhole covers, drainpipe joints.

Only the common castings used in day-to-day applications have been mentioned. The process has a very wide range of applications in engineering.

4.9 Casting treatment

Castings that do not require additional processing can be used as they are. However, most castings that will undergo mechanical loads during use or require further processing, such as machining or metal cutting, need heat treatment. To achieve this, annealing furnaces are commonly found in foundries, where castings are annealed before use or further processing.

4.9.1 Purpose of annealing in casting treatment

- Softens hard steel, making it easier to machine or cold-work>.
- Refines the grain structure, improving ductility and workability.
- Relieves internal stresses developed during solidification and cooling.

1. Annealing process
   i. **Heating phase:**
      - The casting is heated slightly above the AC's critical temperature.
      - It is maintained at this temperature until uniform heat distribution is achieved.
   ii. **Soaking phase:**
      - The casting is held at the critical temperature for a duration known as the soaking time.
      - This period depends on casting thickness, typically 45 min per 25 mm of the largest section.
   iii. **Cooling phase:**
      - The casting is slowly cooled inside the furnace, ensuring uniform temperature reduction in both the surface and the inner core.
      - Slow cooling transforms austenite into softer microstructures, enhancing machinability.

2. Full annealing
   - This process refines the microstructure, reduces hardness, and improves mechanical properties.
   - It ensures uniformity across the casting, making it easier to machine and process further.

By carefully controlling temperature, soaking time, and cooling rate, annealing enhances the quality and durability of castings, ensuring optimal performance in industrial applications.

4.10 Inspection of castings

Inspection of casting can be done in two ways: (a) destructive testing and (b) non-destructive testing. Here we will discuss some non-destructive methods of inspection.

i) **Visual inspection:** Visual inspection refers to the observation of a product with the naked eye. Through this observation, we can identify the following aspects of a product without the use of any instruments.
   a. Roughness of the surfaces
   b. Defect due to mold shift

c. Omission of cores

d. Surface cracks, among other issues, can be detected visually.

ii) **Magnetic particle inspection:** Magnetic particle inspection is performed to check for small voids in castings. This test is generally conducted on ferromagnetic castings. The workpiece is placed in a magnetic field, and particles of ferromagnetic material are spread over the workpiece. These particles accumulate near surface discontinuities.

iii) **Pressure test:** this test is conducted on castings. In this test, the casting is filled with compressed air and submerged in a soap solution. If there is any leakage, it is indicated by bubbles.

iv) **Dye-penetration inspection:** This test is conducted for non-magnetic castings. It identifies surface-visible defects. In this method of testing, the casting is painted or sprayed with fluorescent material. Afterward, the surface is wiped clean, dried, and then viewed in darkness. The defects or discontinuities on the surface become discernible.

v) **Radiographic examination:** this method is used only for those portions that cannot be seen with the naked eye. X-rays and gamma rays are used in this inspection. When the workpiece is placed in the path of X-rays, voids, non-metallic inclusions, porosity, and cracks can be detected on the exposed film.

vi) **Ultrasonic inspection:** In ultrasonic inspection, an ultrasonic signal is sent through the casting. These signals are readily transmitted through a homogeneous medium. If there is a defect or discontinuity in the workpiece, the signal will reflect back and be detected by an ultrasonic detector. The time interval between sending a signal and receiving its reflection determines the location of the discontinuity.

4.11 Quality control of casting

Quality control is concerned with all the functions that affect product quality to ensure that the products achieve the level of quality demanded by the customer. The quality control department determines the required quality level and ensures that the product consistently maintains this standard. Better products result in an increase in market share. Statistical quality control is a tool used to manage quality based on inspection. Recent trends emphasize preventing defects from occurring. Inspections are conducted to assess the quality of an entire population by examining a random sample. A control chart is used to analyze process trends, enabling appropriate action to be taken before the process goes out of control. Recently, the concept of zero defects has emerged. It seeks the voluntary participation of workers in undertaking responsibility for the quality of the task at hand. This approach promotes a high level of error-free work performance. X (mean), R (range) charts, P (percent defective/fraction defective) charts, and C (number of defects in a sample) charts are examples of the tools used to control product quality. The X-bar chart plots the mean of the sample drawn on a chart. All these charts can be used to control the quality of casting.

Questions

Long answer-type questions

1. Discuss the die-casting process, including the differences between hot chamber and cold chamber die-casting. **(IES 2015)**
2. Explain the metal casting process in detail and discuss its various applications in different industries. **(GATE 2016)**
3. Explain the importance of pattern-making in casting. Discuss the different types of patterns used in metal casting. **(IES 2016)**
4. Describe the sand-casting process and explain its advantages and limitations in industrial applications. **(IES 2017)**
5. What are the materials commonly used for making patterns in metal casting, and how do they affect casting quality? **(GATE 2017)**
6. Explain the investment casting process and its advantages in precision casting applications. **(GATE 2017)**
7. Discuss the key steps involved in making a sand casting, highlighting the significance of each step. **(GATE 2018)**
8. Describe the function of a gating system in the metal casting process and explain the different types of gating systems. **(GATE 2019)**
9. Explain the common types of casting defects and their causes. **(GATE 2020)**
10. Describe the functioning of a cupola furnace in the metal casting process. **(IES 2018)**

Short answer-type questions

1. Define the split pattern and its purpose. **(IES 2015)**
2. Why is pattern-making important in the casting process? **(IES 2016)**
3. What is the metal casting process, and where is it primarily used? **(GATE 2016)**
4. List two advantages of sand casting. **(IES 2017)**
5. What materials are commonly used for making patterns in the casting process? **(GATE 2017)**
6. What is the purpose of a gating system in casting? **(IES 2017)**
7. What are the key steps involved in sand casting? **(GATE 2018)**
8. Name two common casting defects and their causes. **(IES 2018)**
9. What are the types of allowances provided to patterns, and why are they necessary? **(GATE 2019)**
10. Differentiate between the hot chamber and cold chamber die casting processes. **(GATE 2020)**

Objective questions

1. Which of the following is a key application of metal casting processes? **(GATE 2020)**
   (a) Manufacturing of solid-state electronic devices.
   (b) Large-scale production of complex components with intricate geometries.
   (c) Fabrication of thin films for optical devices.
   (d) Creating welded joints in high-stress components.
2. Sand casting is commonly used for producing parts with which characteristics? **(IES 2019)**
   (a) High precision and smooth surfaces
   (b) Large and simple geometries
   (c) Lightweight materials
   (d) High tolerance levels

3. Which of the following is the correct sequence of steps involved in sand casting?   **(GATE 2021)**
    (a) Pouring, Solidification, Molding, Pattern Making
    (b) Molding, Pattern Making, Pouring, Solidification
    (c) Pattern Making, Molding, Pouring, Solidification
    (d) Solidification, Pouring, Molding, Pattern Making
4. In pattern-making for sand casting, what is the purpose of the pattern?   **(IES 2020)**
    (a) To create a mold cavity for the molten metal.
    (b) To provide a heat-resistant surface for casting.
    (c) To remove impurities from the metal during casting.
    (d) To cool the molten metal rapidly.
5. Which material is commonly used for making patterns in sand casting?   **(GATE 2019)**
    (a) Wood
    (b) Stainless steel
    (c) Aluminum
    (d) Plastic
6. Which type of pattern is made in multiple parts to facilitate easier removal from the sand mold?
    (IES 2017)
    (a) Solid pattern
    (b) Split pattern
    (c) Skeleton pattern
    (d) Cope and drag pattern
7. What is the purpose of adding pattern allowances in sand casting?   **(GATE 2022)**
    (a) To account for shrinkage during solidification.
    (b) To reduce the overall weight of the pattern.
    (c) To increase the cooling rate of the casting.
    (d) To improve the surface finish of the casting.
8. Which of the following is an essential characteristic of molding sand used in metal casting?   **(IES 2018)**
    (a) High porosity
    (b) High strength
    (c) Low permeability
    (d) High thermal expansion
9. In the die-casting process, which of the following is true for the **hot chamber** process?   **(GATE 2021)**
    (a) The metal is melted in an external furnace.
    (b) It is suitable for metals with high melting points.
    (c) The die is preheated before injection.
    (d) The molten metal is injected into the die using a plunger submerged in the metal.
10. In centrifugal casting, what is the role of rotational motion?   **(IES 2019)**
    (a) To distribute molten metal evenly in the mold
    (b) To control the cooling rate of the casting
    (c) To ensure that the material does not solidify prematurely
    (d) To create a vacuum inside the mold
11. In the cupola furnace, which of the following is used as the primary fuel for melting cast iron?
    (GATE 2020)
    (a) Coke
    (b) Coal
    (c) Gasoline
    (d) Charcoal

12. Which of the following is a common casting defect caused by trapped air during the pouring of molten metal? **(IES 2021)**
    (a) Shrinkage
    (b) Porosity
    (c) Cold shut
    (d) Misrun

13. The primary function of the gating system in casting is to: **(GATE 2019)**
    (a) Cool the molten metal after pouring
    (b) Remove impurities from the molten metal.
    (c) Control the flow of molten metal into the mold cavity
    (d) Solidify the molten metal quickly.

14. Which of the following is a common post-casting treatment used to relieve internal stresses in castings? **(IES 2020)**
    (a) Annealing
    (b) Quenching
    (c) Case hardening
    (d) Normalizing

15. Which of the following is true about machine molding compared to hand molding? **(NET 2020)**
    (a) Machine molding is slower and less accurate.
    (b) Machine molding can achieve better consistency and uniformity.
    (c) Machine molding requires more manpower than hand molding
    (d) Machine molding is only used for very small castings.

Answers

1(B)   2(B)   3(C)   4(A)   5(A)   6(B)   7(A)   8(B)   9(D)   10(A)   11(A)   12(B)   13(C)   14(A)   15(B)

Chapter 5
Welding processes

5.1 Importance and basic concepts of welding

Welding is a fabrication technique used to join materials, primarily metals or thermoplastics, by fusing them together. This is typically achieved by melting the workpieces and introducing a filler material to create a molten weld pool, which solidifies into a strong joint. In some cases, pressure is applied along with heat or independently to complete the weld.

As a crucial process across industries of all sizes, welding serves as a primary method for manufacturing and repairing metal products. It is efficient, cost-effective, and highly reliable. Notably, welding is the only joining method tested in space and is suitable for applications on land, underwater, and in space. Its fundamental role in engineering and construction supports technological advancements and industrial development, emphasizing its significance in the modern world.

Welding is crucial for several reasons across various industries due to its unique capabilities and advantages:

1. **Structural integrity:** Welding provides strong, durable joints that can withstand high stress and pressure, making it essential for constructing buildings, bridges, and other infrastructure.
2. **Versatility:** Welding can join a wide range of materials, including different types of metals and thermoplastics, in various shapes and thicknesses.
3. **Economic efficiency:** Welding is cost-effective, offering a reliable and long-lasting solution for joining materials. It often reduces the need for additional materials, such as fasteners or adhesives.
4. **Wide range of applications:** Welding is used in numerous industries, including automotive, aerospace, construction, shipbuilding, and manufacturing, playing a critical role in the production and repair of equipment, vehicles, and structures.
5. **Innovation and advancements:** The development of advanced welding techniques, such as laser and electron beam welding, enables precision work in high-tech industries, including electronics and medical device manufacturing.
6. **Adaptability to extreme conditions:** Welding can be performed in various environments, including underwater, in space, and in challenging industrial settings, making it indispensable for specific applications such as underwater pipelines and space station construction.
7. **Repair and maintenance:** Welding is vital for repairing and maintaining equipment and structures, extending their lifespans and ensuring safety and functionality.
8. **Efficiency and speed:** Welding processes are generally faster than other joining methods, allowing for quicker production and assembly times.

https://doi.org/10.1515/9783112205891-005

9. **Environmental benefits:** Modern welding techniques are more environmentally friendly, reducing waste and energy consumption compared to other fabrication methods.

5.2 Classification of welding process

Welding processes can be categorized in multiple ways, with one of the most logical classifications based on the energy sources used for coalescence. Broadly, welding processes are classified into the following categories:

(a) **Fusion welding**

1. **Gas flame welding**
   - Oxyacetylene welding
   - Oxyhydrogen welding

2. **Electric arc welding**
   - Carbon arc welding
   - Shielded metal arc welding
   - Submerged arc welding
   - Tungsten arc welding
   - Metal inert gas (MIG) welding
   - Plasma arc welding
   - Atomic hydrogen welding

3. **Radiant energy welding**
   - Electron beam welding
   - Laser beam welding

4. **Electric resistance welding**
   - Spot welding
   - Seam welding
   - Projection welding
   - Resistance butt welding
   - Flash welding
   - Percussion welding
   - Electroslag welding

The following processes fall under solid-state (cold) welding:

(b) **Pressure welding**
   - Friction welding
   - Ultrasonic welding
   - Explosive welding
   - Forge and diffusive welding

(c) **Thermochemical welding**
- – Thermit welding
- – Atomic hydrogen welding

Among these, carbon arc welding and atomic hydrogen welding are now rarely used in industrial applications.

5.3 Fusion welding

Fusion welding consists of processes that join metals by heating the contact areas above their melting points, allowing them to flow and fuse together. In most cases, filler materials are added to ensure a properly filled joint. To achieve high-quality welds, all fusion welding processes require four essential elements:

1. **Energy source:** Provides the necessary heat for fusion, such as a gas flame, an electric arc, or electric resistance.
2. **Surface preparation:** Removes contaminants such as organic residues and oxide films to ensure a clean welding surface.
3. **Atmospheric protection:** Shields the molten metal from contamination by using gases such as argon, helium, or carbon dioxide, or through smoke generated by flux coatings.
4. **Weld metallurgy control:** Involves techniques such as preheating and post-weld heat treatment to manage the weld's structural integrity and mechanical properties.

5.3.1 Gas welding (oxyacetylene welding)

Oxyacetylene welding is a fusion welding process that joins metals using heat generated from the combustion of a high-temperature gas flame. This intense heat raises the temperature of the metal edges, melting and fusing them together. A filler metal may be added to the molten area to fill the gap between the workpieces.

Different oxy-fuel combinations, such as oxygen-acetylene ($O_2 + C_2H_2$) or oxygen-hydrogen ($O_2 + H_2$) with coal gas, produce various types of heating flames. Among these, oxyacetylene welding is the most widely used gas welding process.

In this process, a welding torch precisely mixes oxygen and acetylene in the correct ratio, directing the flame onto the metal surfaces to be joined. The molten edges fuse together and, upon cooling, form a strong bond. A filler rod, often coated with flux powder, is typically added to the joint, melting into the molten metal from the parent material to enhance the weld quality.

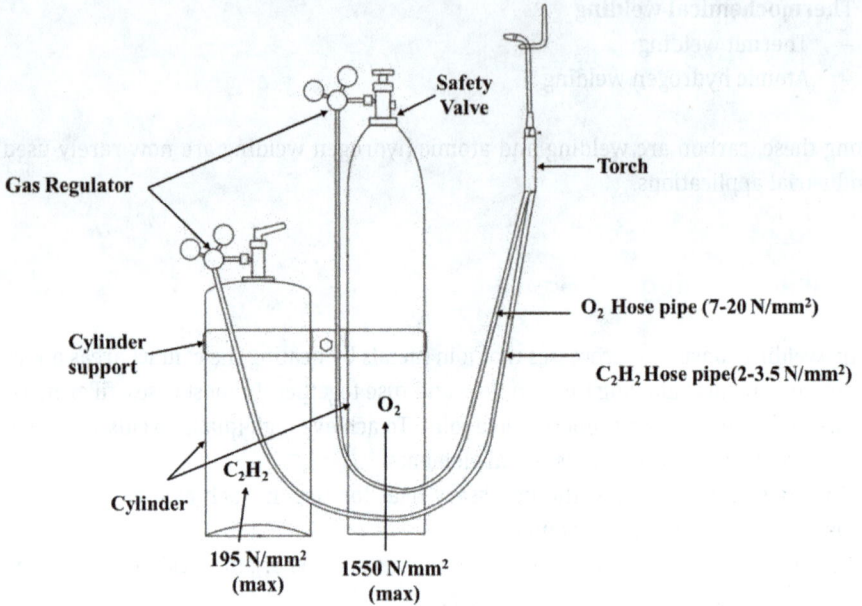

Figure 5.1: Oxyacetylene gas welding equipment.

Acetylene is preferred due to its high flame temperature when mixed with oxygen, reaching between 2,100 °C and 3,500 °C. This temperature significantly exceeds the melting point of most metals, enabling rapid and localized melting, which is essential for effective welding. Additionally, the oxyacetylene flame is widely used for cutting ferrous metals.

Oxyacetylene welding and cutting are especially popular in maintenance work due to the ease of flame control, the affordability of gases, the portability of equipment, and the overall safety of the process.

Combustion of gas and heat liberated.

First stage

$$C_2H_2 + O_2 \rightarrow 2\,CO + H_2 + 448\,kJ/mol$$

Second stage

$$2\,CO + H_2 + 3O \rightarrow 2\,CO_2 + H_2O + 812\,kJ/mol.$$

Total heat liberated by combustion = 448 + 812 = 1,260 kJ/mol of acetylene.

1. **Gas welding equipment:** Figure 5.1 shows a detailed schematic diagram of all major equipment.
   1) Oxygen gas cylinder
   2) Acetylene gas cylinder

3) Welding torch
4) Pressure regulator
5) Hose pipe
6) Stop valve
7) Pressure gauge

2. Types of flame

For effective and efficient welding, it is essential to mix the gases in the correct proportions. Figure 5.2 illustrates the basic design of a gas welding flame.

$$C_2H_2 + O_2 = 2\ CO + H_2$$

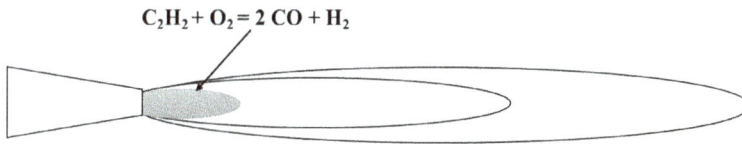

Figure 5.2: Gas welding flame.

Based on the volume ratio of O_2 and C_2H_2, the flame can be classified as follows:

1. Neutral flame

A neutral flame is formed when oxygen (O_2) and acetylene (C_2H_2) are mixed in equal volumes. It features a distinct inner cone with a light blue hue. This flame neither alters the molten metal nor causes oxidation or carburization, making it ideal for welding materials such as mild steel, cast iron, aluminum, and stainless steel. The temperature of a neutral flame can reach up to 3,232 °C.

2. Oxidizing flame

An oxidizing flame is produced by increasing the oxygen content in the gas mixture. It has a small, sharp white inner cone that is shorter and more pointed than a neutral flame. With a high temperature of 3,482 °C, this flame is generally not used for most welding applications, except for welding brass.

3. Reducing flame

A carburizing flame is formed when the acetylene (C_2H_2) supply exceeds the oxygen supply. It reaches an approximate temperature of 3,150 °C and has a longer outer envelope compared to a neutral flame. This flame is primarily used for welding lead. The highest temperature is located at the tip of the inner cone. A distinctive feature of carburizing flames is the feather-like or brush-like extension adjacent to the inner cone.

Figure 5.3 depicts all three types of flames and their corresponding temperatures.

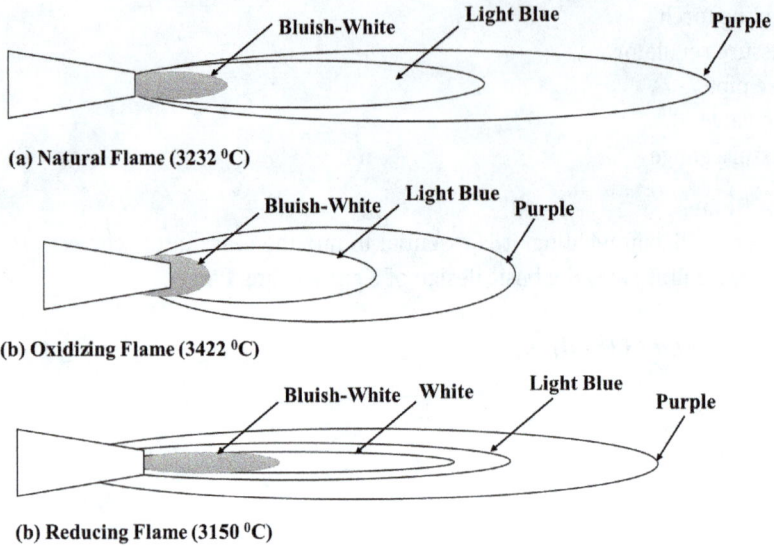

(a) Natural Flame (3232 ⁰C)

(b) Oxidizing Flame (3422 ⁰C)

(b) Reducing Flame (3150 ⁰C)

Figure 5.3: Types of flames.

5.3.2 Electric arc welding

In electric arc welding, the metal pieces to be joined are heated to their melting point by generating an electric arc between them. This process creates a molten metal pool, which solidifies to form a strong welded joint. In some cases, a filler metal is added by melting a wire to enhance the weld.

Electric arc welding is categorized into two types:
1. Metal arc welding
2. Carbon arc welding (obsolete)

1. Metal arc welding

In metal arc welding, an electric arc is sustained between the electrode and the workpiece, which serve as the two terminals. The electrode can be either bare or coated. A bare electrode has the same composition as the parent metal, whereas a coated electrode contains flux material to prevent surface oxidation.

To generate the required heat, the arc is initiated by briefly touching the electrode to the workpiece and then withdrawing it to an appropriate distance. This arc generates intense heat, melting both the electrode tip and the workpiece. As the electrode melts, molten metal droplets are transferred through the arc and deposited along the joint. The coated electrode's flux burns, producing a protective gas shield around the arc, which prevents atmospheric contamination of the molten weld metal. Figure 5.4 illustrates a schematic diagram of an electric arc welding setup.

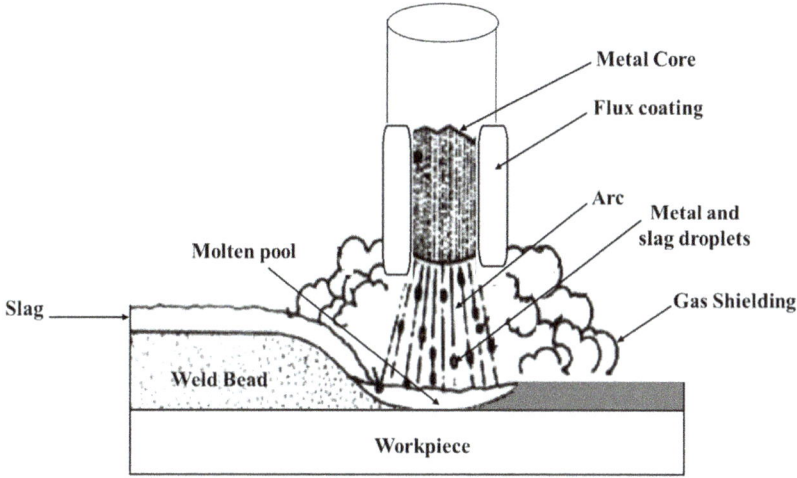

Figure 5.4: Electric arc welding.

a) Arc welding power source

In arc welding, both AC and DC power sources can be used. In DC welding

Polarity in arc welding:

When AC (Alternating Current) is used in welding, polarity is not fixed at any terminal and alternates with each cycle, resulting in equal heat distribution at both poles, however, in DC (Direct Current) welding, the polarity remains fixed, where the workpiece serves as one terminal and the electrode as the other. The workpiece (job) serves as one terminal, while the electrode acts as the other. The heat distribution in DC welding is uneven, with two-thirds (2/3) of the total heat generated at the positive terminal and one-third (1/3) at the negative terminal.

DC welding has two types of polarity, as mentioned in Figure 5.5.

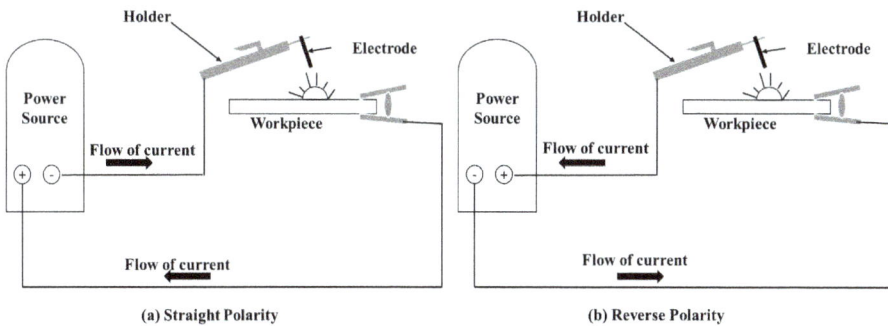

Figure 5.5: Polarity in electric arc welding.

- **DC straight polarity (DCSP):** Electrode negative (-), workpiece positive (+). This polarity is used in the welding of thick materials due to the significant requirement of heat on the plate.
- **DC reverse polarity (DCRP):** Electrode positive (+), workpiece negative (-). This polarity is used in the welding of thin materials due to the reduced requirement of heat in the welding zone.

b) Types of electrodes in arc welding

In arc welding, electrodes can be broadly classified into two main types: consumable and non-consumable electrodes. These categories are further divided based on their specific characteristics and applications.

1. **Consumable electrodes**

Consumable electrodes are designed to melt and become part of the weld, providing filler material for the weld joint.

Types of consumable electrodes:
- **Bare electrodes:**
    - Made of plain wire without any coating.
    - Used for welding similar metals.
    - Offer less protection against oxidation and contamination.
- **Coated electrodes:** Covered with a flux coating.
    - Provide better protection against oxidation and contamination.
    - Improve arc stability and weld quality.
    - Common types include:
        - **Cellulose-coated electrodes:** Produce a gas shield and are used for deep penetration welds.
        - **Rutile-coated electrodes:** Contain titanium dioxide (rutile) and offer smooth welding and easy slag removal.
        - **Basic coated electrodes** Contain calcium carbonate and other minerals, providing good toughness and crack resistance.

2. **Non-consumable electrodes**

Non-consumable electrodes do not melt or become part of the weld; they are used primarily to create the arc.

Types of non-consumable electrodes:
- **Carbon electrodes:**
    - Made of carbon or graphite.
    - Used in applications requiring a high heat input.
    - Obsolete for most modern welding processes, but it is still used in specific applications like gouging.
- **Tungsten electrodes:**
    - Made of tungsten, with or without alloying elements.

- Used primarily in gas tungsten arc welding (GTAW or TIG).
- Provide high-temperature resistance and conductivity.
- **Types include:**
  - **Pure tungsten electrodes:** Contain 99.5% tungsten.
  - **Thoriated tungsten electrodes (ThO_2):** Contain thorium oxide, offering improved arc stability and longevity.
  - **Ceriated tungsten electrodes (CeO_2):** Contain cerium oxide, providing easier arc starting.
  - **Lanthanated tungsten electrodes (La_2O_3):** Contain lanthanum oxide, offering good arc stability and low erosion rates.

Each type of electrode has specific properties and applications, making them suitable for different welding tasks and materials.

5.4 Resistance welding

Resistance welding is a process used to join two metal pieces by applying both mechanical pressure and heat. The heat is generated due to the electrical resistance of the metal pieces as an electric current flows through them. In this method, the metal pieces are held together while a high electric current is passed through the joint. The resistance to current flow causes the temperature at the interface to rise to the fusion point. At this moment, slight mechanical pressure is applied to complete the weld.

The current used in resistance welding typically ranges from 3,000 to 100,000 amperes for a fraction of a second, with a voltage between 1 and 25 volts. The heat generated in resistance welding is given by the formula:

$$H = I^2 \, RT$$

where
- **H** = Heat generated (Joules)
- **I** = Current (Amperes)
- **R** = Electrical resistance of the material (ohms)
- **T** = Time of current flow (seconds)

Resistance welding can be further classified.

5.4.1 Resistance spot welding

Spot welding is a resistance welding process in which overlapping metal sheets are clamped between two copper electrodes. These electrodes focus the welding current at a specific point, generating heat to melt the interface while applying pressure to form a solid weld. Figure 5.6 shows the basic details of resistance spot welding.

Key features of spot welding:
1. **Current range**: Typically, between 3,000 and 10,000 amperes.
2. **Weld zone temperature:** Reaches between 815 °C and 930 °C.
3. **Electrode material:** Usually made of copper-based alloys.
4. **Weld nugget size:** Generally, 6–10 mm in diameter.

Advantages:
–   High welding rates
–   Minimal fumes
–   Cost-effective process
–   Easily automated
–   No need for filler materials
–   Low distortion in welded parts

Disadvantages:
–   High initial equipment cost
–   Discontinuous welds have lower strength
–   Limited to welding sheet thicknesses up to 1/4" (6 mm)

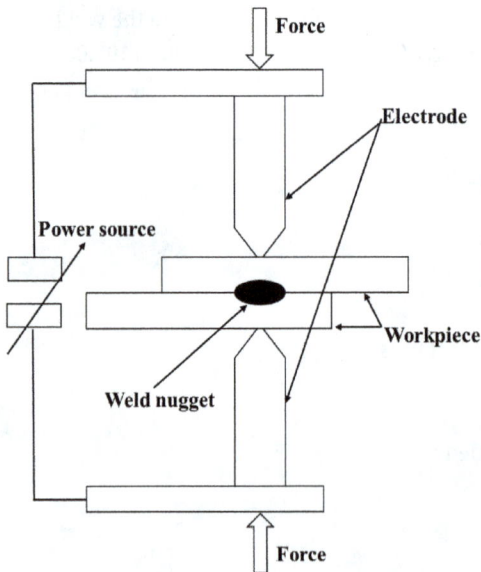

Figure 5.6: Resistance spot welding.

5.4.2 Resistance seam welding

Resistance seam welding is a variation of spot welding that uses continuously rotating circular electrodes instead of stationary tip electrodes. This process produces a continuous, airtight seam by creating a series of overlapping spot welds along the joint. Figure 5.7 depicts the fundamental setup of resistance seam welding for ease of understanding. It is commonly used in applications requiring leak-proof and high-strength welds, such as automotive fuel tanks, pipe welding, and aerospace industries.

Key features of resistance seam welding:
– **Continuous weld:** Produces a leak-proof, airtight seam.
– **Electrode type:** Uses circular, rotating copper electrodes instead of pointed tips.
– **Weld formation:** Creates a series of overlapping spot welds to form a continuous joint.
– **Current and pressure:** A controlled flow of electric current and electrode pressure ensure uniformity.
– **Typical applications:** Used in fuel tanks, radiators, pipelines, and sealed containers.

Advantages:
– **Airtight and watertight welds:** Suitable for applications requiring sealed joints.
– **High production rate:** Faster than individual spot welding.
– **Automatable process:** Can be easily integrated into mass-production lines.
– **No need for additional materials:** Works without filler metal or additional flux.
– **Minimal distortion:** Produces uniform and consistent welds with low heat input.

Disadvantages:
– **High equipment cost:** Requires specialized seam-welding machines, which are expensive.
– **Limited to thin sheets:** Typically used for materials up to 3 mm thick; unsuitable for thicker materials.
– **Electrode wear and maintenance:** Continuous use causes electrode wear, requiring frequent grinding or replacement.
– **Limited weld geometry:** Not suitable for complex shapes; mainly used for straight or gently curved seams.

5.4.3 Resistance projection welding

Resistance Projection Welding (RPW) is a resistance welding process in which the weld is localized at predetermined points by small projections or embossments on one or both workpieces. These projections concentrate the welding current, generate

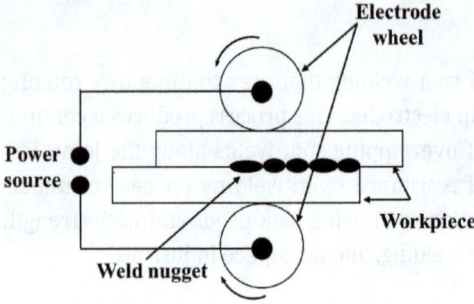

Figure 5.7: Resistance seam welding.

heat, and form a strong joint when pressure is applied. Figure 5.8 visually depicts the working mechanism. It is commonly used for welding fasteners, nuts, studs, and sheet metal assemblies.

Key features:
- **Localized heat generation:** Heat is concentrated at predefined projections on the workpiece.
- **Multiple welds in one cycle:** Several projections can be welded simultaneously, increasing efficiency.
- **Electrode life extension:** Reduced electrode wear compared to spot welding, as the current flows through small contact areas.
- **Suitable for various materials:** Works well with low-carbon steels, stainless steel, aluminum, and coated metals.
- **Common applications:** Used for fasteners, electrical components, sheet metal parts, and automotive structures.

Advantages:
- **Improved weld strength:** Concentrated heat results in strong, uniform welds.
- **Efficient and fast:** Allows multiple projections to be welded in a single operation.
- **Reduced electrode wear:** Less deformation of electrodes compared to spot welding.
- **Consistent weld quality:** Predefined projection locations ensure precise, repeatable welds.
- **No filler material required:** Like other resistance welding processes, it does not require filler material.

Limitations:
– **Requires preformed projections:** Workpieces must be designed with projections, increasing preparation time.
– **Limited to certain thicknesses:** Generally used for thin to medium-thickness materials (not ideal for very thick parts).
– **Electrode alignment critical:** Proper electrode positioning is essential for uniform weld quality.
– **Higher initial cost:** Requires specialized machinery and tooling for mass production.

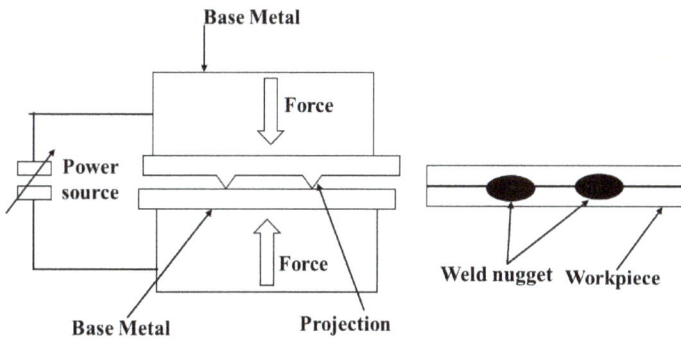

Figure 5.8: Resistance projection welding.

5.4.4 Flash welding (FW)

Flash welding (FW) is a resistance welding technique used to join metal components by applying high electrical current and pressure. The process begins with flashing, where an electric arc forms between the workpieces, causing localized melting. Once the desired temperature is reached, the parts are forcibly pressed together, creating a strong bond without requiring filler material.

In flash welding, the ends of rods, tubes, or sheets are heated and fused by an electric arc before being pressed together to complete the weld. The workpieces are held in electrode clamps, with one clamp fixed and the other movable. This method enables rapid joining – typically within one minute – making it efficient for large and complex parts. Figure 5.9 illustrates a fundamental understanding of flash welding. After welding, annealing is often performed to enhance the weld's toughness. Flash welding is suitable for various materials, including steel, aluminum alloys, copper alloys, magnesium alloys, and nickel alloys. It is widely used in manufacturing thick pipes, band saw ends, frames, and aircraft landing gear.

Key features:
- **No filler material required:** The weld is formed purely by heat and pressure.
- **High electrical current:** Generates intense heat, melting the interface between workpieces.
- **Flash formation:** Sparks and molten metal are ejected as the surfaces heat up and burn away impurities.
- **Rapid process:** Welding takes only a few seconds to a few minutes, depending on the material thickness.
- **Common applications:** Used in rail welding, automotive components, and structural steel fabrication.

Advantages:
- **High-quality welds:** Produces strong, defect-free joints with minimal impurities.
- **No need for filler metal or flux:** Reduces material costs and simplifies the process.
- **Fast and efficient:** Ideal for mass production and industrial applications.
- **Good for large cross-sections:** Suitable for welding thick materials such as rails, pipes, and rods.
- **Automatable process:** Can be easily integrated into production lines.

Limitations:
- **High initial equipment cost:** Requires specialized flash-welding machines.
- **Material restrictions:** Best suited for ferrous metals; not ideal for non-metals or certain alloys.
- **Excessive flash formation:** Requires post-weld trimming to remove excess metal.
- **Limited precision:** Not ideal for delicate or precision welding applications.

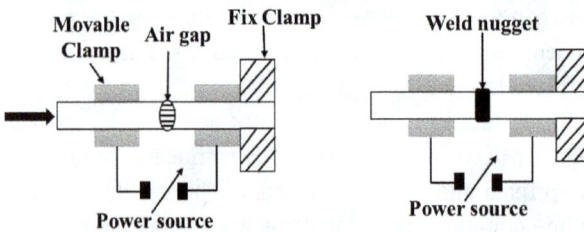

Figure 5.9: Flash welding.

5.4.5 Resistance butt welding

Figure 5.10 depicts a schematic diagram of resistance butt welding. Resistance Butt Welding (BW) is a resistance welding process used to join metal parts end-to-end by applying electric current and pressure. In this method, the ends of wires or rods are pressed together, and heat is generated through the electrical resistance at the contact point. Once the metal reaches a plastic state, forging pressure is applied to create a strong and seamless joint.

Unlike flash welding, resistance butt welding does not produce an arc or flashing, ensuring minimal material loss during the process. This method is highly efficient, clean, and ideal for welding small parts, making it suitable for mass production applications.

Key features:
–   **No filler material required:** The weld is formed using only heat and pressure.
–   **No flashing or sparks:** Unlike flash welding, the surfaces are heated and pressed together smoothly.
–   **High-current, low-voltage process:** A strong electric current is passed through the interface to generate heat.
–   **Used for end-to-end joints:** Ideal for welding rods, wires, pipes, and bars.
–   **Common applications:** Used in rail welding, wire manufacturing, chain link production, and automotive parts.

Advantages:
–   **Smooth and strong joints:** Produces clean welds without excessive flash.
–   **High efficiency:** The process is fast and energy-efficient, making it ideal for mass production.
–   **No filler or flux needed:** Reduces material costs and simplifies the process.
–   **Good for conductive metals:** Works well with copper, aluminum, and steel.
–   **Consistent weld quality:** Automated control ensures uniform results.

Limitations:
–   **Limited to similar metals:** Works best with materials of similar compositions and thicknesses.
–   **High initial equipment cost:** Requires specialized welding machines.
–   **Not suitable for thick sections:** More effective for small to medium-sized components.

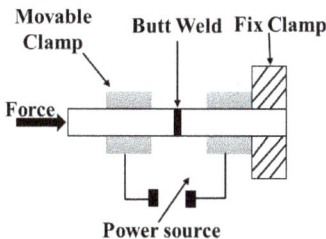

Figure 5.10: Resistance butt welding.

5.4.6 Percussion welding

Percussion welding (PW), illustrated in Figure 5.11, is a resistance welding process that joins metal parts by applying a brief, high-intensity electric discharge along with a rapid mechanical impact (percussive force). The heat generated at the interface brings the metal to a plastic state, and the sudden impact forces the surfaces together, creating a strong joint.

In this process, the materials are joined simultaneously across the entire contact area, utilizing heat from a rapid discharge of electrical energy. Pressure is applied percussively – either during or immediately after the electrical discharge – to ensure a solid and reliable bond.

Key features:
- **Rapid process:** The welding occurs in milliseconds due to the high-intensity electrical pulse.
- **Percussive impact:** A mechanical force is applied immediately after the heat is generated to forge the joint.
- **Minimal heat-affected zone (HAZ):** Since the heat is highly localized, the surrounding material remains unaffected.
- **No filler material or flux required:** The weld is formed purely by heat and impact, making it a clean process.
- **Suitable for dissimilar metals:** Effective for joining highly conductive metals, such as copper to steel or aluminum to copper.

Advantages:
- **Extremely fast welding:** The process is completed in milliseconds, enhancing production efficiency.
- **Strong and clean joints:** Produce high-strength welds with minimal contamination.
- **Low heat input:** Reduces the risk of material distortion and improves weld quality.
- **Good for dissimilar metals:** Suitable for materials with different melting points and electrical conductivities.

Limitations:
- **Requires specialized equipment:** Needs a high-energy power source and precise control systems.
- **Limited to specific applications:** Primarily used for electrical contacts, battery terminals, and fine metal components.
- **Not ideal for large structures:** Works best for small, precision components rather than large assemblies.

Common applications:
– Welding electrical contacts (e.g., copper to steel).
– Battery terminals and small electronic components.
– Joining thin wires and metal foils.
– Manufacturing precision components in the aerospace and automotive industries

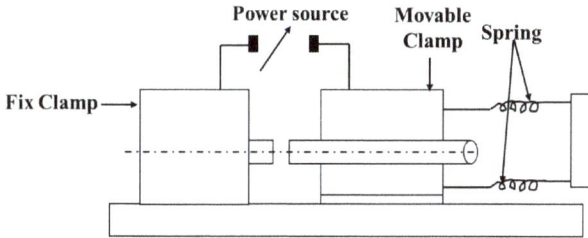

Figure 5.11: Percussion welding.

5.5 Soldering and brazing

5.5.1 Soldering

Soldering is a low-temperature metal joining process in which a filler metal (solder) is melted to bond metal parts without melting the base materials. The solder, typically made of tin-lead or other metal alloys, flows into the joint through capillary action, creating a strong electrical and mechanical connection.

The process involves heating the joint to a temperature below 450 °C, allowing the solder to melt and flow between closely fitted surfaces. The heat also activates the flux, which cleans the metal surfaces and enhances bonding. Figure 5.12 depicts the basic workings of the soldering process. Manual soldering, particularly for critical electronic equipment and components, demands a high level of skill. In contrast, automated soldering requires minimal operator expertise, as process variables are preset. The quality of the joint is determined by machine settings, process control, and inspection. While the lap joint is the most commonly used joint, specialized joint types are employed for soldering electronic components.

Key steps for successful soldering:
i.   **Proper fit-up:** Ensure the metal parts are closely fitted.
ii. **Surface preparation:** Clean the surfaces to remove oxides and contaminants.
iii. **Flux application:** Apply flux to prevent oxidation and to promote adhesion.
iv. **Heating and soldering:** Assemble the parts, apply heat, and melt the solder.
v.   **Flux residue removal:** Clean off excess flux after the joint has cooled.

Key features:
- **Low-temperature process:** Soldering occurs below 450 °C (842 °F), preventing damage to sensitive components.
- **Capillary action:** The molten solder spreads between closely fitted metal surfaces without external pressure.
- **No fusion of base metals:** Unlike welding, the workpieces remain intact; only the filler metal melts.
- **Use of flux:** Flux is applied to remove oxides and to ensure proper wetting of the solder.

Advantages:
- **Low heat input:** Reduces the risk of warping or damaging heat-sensitive components.
- **Joins dissimilar metals:** Suitable for metals like copper, brass, aluminum, and stainless steel.
- **Good electrical conductivity:** Commonly used in electronic circuit boards and electrical connections.
- **Clean and precise joints:** Produces smooth, neat joints with minimal finishing required.
- **Cost-effective:** Requires inexpensive equipment and materials compared to welding.

Limitations:
- **Lower strength:** Soldered joints are weaker than welded or brazed joints and are not suitable for high-load applications.
- **Not heat-resistant:** Soldered joints may weaken at high temperatures.
- **Requires clean surfaces:** Any oxidation or contamination can prevent proper bonding.

Common applications: This process is widely used in electronics, plumbing, jewelry, and automotive applications due to its efficiency, precision, and ability to join dissimilar metals.
- This method is suitable for joining metal components of varying thicknesses, ranging from delicate thin films to heavy-duty parts such as bus bars and piping.
- High-cost automated equipment can efficiently produce multiple high-quality joints at once, significantly lowering the cost per joint.
- While manual soldering is a slower process, it remains a cost-effective option for low-volume production or intricate joint configurations.
- **Electronics and PCB assembly:** Used in circuit boards, wiring, and electrical components.
- **Plumbing and pipe fittings:** For copper and brass pipes in HVAC and plumbing systems.

- **Jewelry and fine metalwork:** Create delicate and precise joints in gold, silver, and brass.
- **Automotive and aerospace:** Used for small electrical connections and heat exchangers.

Soldering methods

i. Dip soldering	v. Induction soldering
ii. Iron soldering	vi. Furnace soldering
iii. Resistance soldering	vii. Infrared soldering
iv. Torch soldering	viii. Ultrasonic soldering

Solders can be classified as:

i. **Soft solder:** A lead-tin alloy with varying compositions
    - 50% Tin, 50% Lead
    - 67% Tin, 33% Lead
    - 33% Tin, 67% Lead

Soft solder melts at temperatures below 350 °C, and the process is known as soft soldering.

ii. **Hard solder:** A copper-zinc alloy that melts at temperatures above 600 °C. The process, commonly referred to as hard soldering, is typically performed manually.

To prevent oxidation of the joint surfaces, fluxes are used. The flux should be lightweight so that it can be easily displaced by the molten metal. Examples of fluxes include zinc chloride, rosin, rosin plus alcohol-based flux, and a mixture of zinc chloride and ammonium chloride.

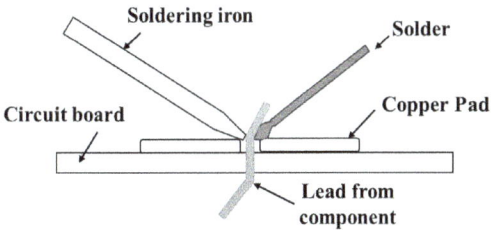

Figure 5.12: Soldering operation.

5.5.2 Brazing

Brazing is a metal-joining process in which a filler metal is heated above 450 °C and flows between closely fitted parts through capillary action, without melting the base metals. The filler metal's melting point is lower than that of the base material but high enough to create a strong bond.

In brazing, the filler metal is applied directly into a groove at the joint rather than relying on capillary action. Unlike brazing, soldering uses a filler metal with a melting point below 450 °C, making it suitable for lower-temperature applications.

Key aspects of brazing:
– The joint is formed through capillary action, ensuring strong and durable bonds.
– The process requires flux to remove oxides and promote proper wetting of the surfaces.
– It provides high strength, corrosion resistance, and the ability to join dissimilar metals.
– Common filler metals include alloys of silver, copper, zinc, aluminum, and nickel.

Key elements for successful brazing include:
1. Joint design: Proper fit and clearance between parts ensure capillary action for effective filler metal flow.
2. Filler metal: A suitable alloy with a melting point above 450 °C but lower than the base metal's melting point.
3. Uniform heating: Ensures even heat distribution for complete bonding without distortion.
4. Protective or reactive cover: Use of flux or a controlled atmosphere to prevent oxidation and ensure a clean joint.

Advantages of brazing:
– **Strong joints:** Produces high-strength, leak-proof joints.
– **Dissimilar metals:** Can join different metals that are difficult to weld.
– **Minimal distortion:** Operates at lower temperatures than welding, thereby reducing warping.
– **Multi-joint capability:** Multiple joints can be brazed simultaneously.
– **Automation-friendly:** Easily adaptable to automated production processes.

Disadvantages of brazing:
– **Lower strength than welding:** Not suitable for high-stress applications.
– **Requires flux:** Oxidation must be controlled to ensure proper bonding.
– **Heat sensitivity:** May not be ideal for materials that are sensitive to high temperatures.
– **Color mismatch:** The filler metal may not match the base metal in appearance.

Common application:
Brazing is widely used in HVAC systems, plumbing, aerospace, automotive, and electronics industries, where strong, reliable, and precise joints are required.

Brazing processes are classified based on their heating methods:
- **Torch brazing:** Uses an oxy-fuel gas torch to weld thin sections (0.25 mm to 6 mm), typically for lap joints.
- **Furnace brazing:** Involves preplacing flux and filler metal before heating in a controlled furnace atmosphere.
- **Induction brazing:** Uses electromagnetic induction to heat and join sections up to 25 mm thick.
- **Resistance brazing:** Employs electric resistance heating, suitable for sheets ranging from 0.1 mm to 12 mm.
- **Dip brazing:** Involves immersing components in a molten salt or metal bath for uniform heating.
- **Infrared brazing:** Utilizes infrared radiation to generate heat for the brazing process.
- **Diffusion brazing:** Depends on the diffusion of elements at the joint interface to create a strong bond.

Additional methods:
i. **Vacuum brazing:** Conducted in a vacuum chamber to prevent oxidation and contamination.
ii. **Laser brazing:** Uses a laser to precisely heat the joint area.
iii. **Ultrasonic brazing:** Employs ultrasonic vibrations to aid in the distribution of the filler metal.

5.6 Welding defects and remedies

Welding defects reduce weld quality and strength and are frequently caused by incorrect parameters, poor technique, or contamination. Preventive measures include following the correct procedure, maintaining heat control, ensuring clean materials, and selecting appropriate electrodes. Regular examination ensures that welds are defect-free. Different welding defects are shown in Figure 5.13, and details are summarized in Table 5.1.

5.7 Welded joint and edge preparation

Proper welded joint and edge preparation are critical for producing robust, defect-free welds. To guarantee good penetration and fusion, select the appropriate joint type

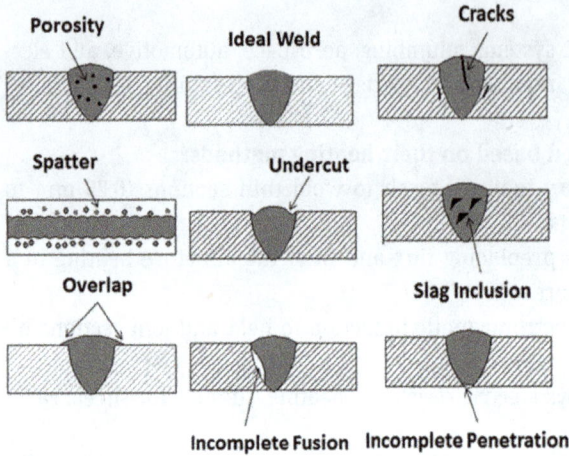

Figure 5.13: Different types of welding defects.

Table 5.1: Summary of weld defects and their remedies.

Defect	Cause	Remedy
Porosity	Gas entrapment during solidification	Use clean, dry electrodes, clean the surface, and maintain proper welding speed and shielding gas flow
Undercut	Excessive current or high travel speed	Reduce welding current, use the correct electrode angle and travel speed, and avoid over-manipulation of the arc
Cracks	Rapid cooling, poor joint design, or residual stresses	Preheat the material; use proper techniques and filler materials; ensure uniform cooling
Incomplete fusion	Low heat input or improper electrode angle	Increase welding current; ensure proper joint preparation and arc control
Slag inclusion	Improper slag removal or poor electrode handling	Clean slag between passes; use appropriate techniques and consumables
Distortion	Uneven heat distribution	Use clamping and fixturing, control heat input, and employ balanced welding sequences
Spatter	High current or incorrect arc length.	Optimize welding current and voltage; use anti-spatter spray if necessary
Overlap	Excessive deposition of weld metal	Maintain the correct electrode angle and travel speed; reduce heat input
Lack of penetration	Low welding current or improper joint preparation.	Increase the current; ensure proper joint design and edge preparation
Burn through	Excessive heat input on thin materials	Use appropriate current settings; weld in short intervals to avoid overheating

(butt, lap, fillet, etc.) and prepare the edges by machining, grinding, or beveling. Clean, well-prepared edges increase weld strength, eliminate flaws, and improve overall weld quality. Different types of weld joints and edge preparations are illustrated in Figure 5.14, with Table 5.2 summarizing common weld joint types , along with their typical applications in structural and fabrication settings. Additionally, Table 5.3 outlines various edge preparation methods, highlighting their descriptions and suitability for different plate thicknesses and welding requirements.

Table 5.2: Summary of weld joints and their applications.

Welded joint type	Description	Applications
Butt joint	Two plates joined edge-to-edge in the same plane	Sheet metal, plates, and pipes
Lap joint	One plate overlaps another, welded along the edges	Thin sheets, automotive applications
Corner joint	Plates meet at right angles to form an L-shape	Frames, sheet metal fabrications
T-joint	One plate is perpendicular to another, forming a "T"	Structural connections, frameworks
Edge joint	Edges of two plates are placed side-by-side and welded	Low-stress applications

Table 5.3: Summary of edges in welding with their applications.

Edge preparation type	Description	Applications
Square edge	Edges are left as is without special preparation	Thin plates, low-strength welds
Single bevel	One edge beveled to form a groove	Thicker plates for full penetration
Double bevel	Both edges beveled, creating a V-shaped groove	Plates requiring full penetration welds
Single V-groove	Both plate edges beveled to form a V-shape	Medium to thick plates
Double V-groove	Beveled on both sides, forming a double V	Very thick plates requiring penetration on both sides
J-groove	One plate edge is curved like a "J"	Thick plates, reduce weld metal usage
U-groove	Both edges curved to form a "U"	Very thick plates, reduce filler metal usage
Flanged edge	Edges flanged to increase the weld area	Thin plates, improves weld strength

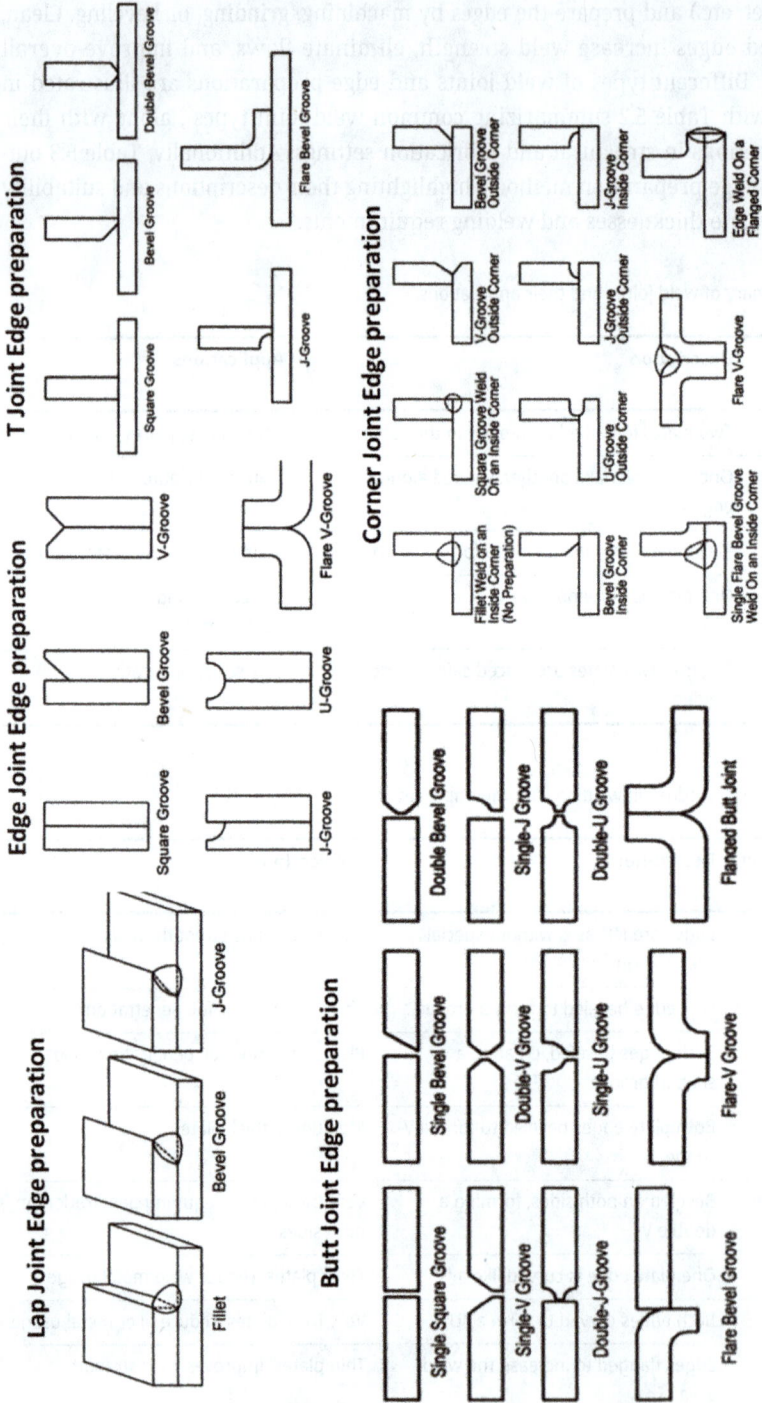

Figure 5.14: Different types of weld joints and edge preparations.

Question

Long answer-type questions

1. Explain the principle of metal arc welding (MAW) and discuss the different types of electrodes used in the process. How do these electrodes affect the weld quality?
2. Discuss the common defects encountered in metal arc welding. Explain the causes and preventive measures for each defect.
3. Describe the various types of metal arc welding techniques, such as shielded metal arc welding (SMAW), gas metal arc welding (GMAW), and flux-cored arc welding (FCAW). Compare their applications, advantages, and limitations.
4. What is the significance of preheating and post-weld heat treatment in metal arc welding? Discuss their effects on the mechanical properties of the welded joint.
5. Explain the safety precautions and equipment required in metal arc welding. How does the choice of personal protective equipment (PPE) impact welder safety?
6. Describe the oxyacetylene gas welding process, including the equipment setup, flame characteristics, and the types of flames used. How does the flame type affect the quality of the weld?
7. Compare and contrast the oxyacetylene welding process with oxy-hydrogen welding. Discuss the advantages, disadvantages, and specific applications of each.
8. Explain the procedure for gas welding of dissimilar metals. What are the challenges associated with welding different metals, and how can they be overcome?
9. Discuss the factors influencing the selection of filler materials in gas welding. How does the choice of filler material affect the mechanical properties of the weld?
10. Explain the role of flux in gas welding. How does flux composition and application method influence the welding process and the final weld quality?
11. Describe the principle and working of spot welding. Discuss the factors that affect weld strength and quality in resistance spot welding.
12. Compare and contrast the different types of resistance welding processes, such as seam welding, projection welding, and flash welding. Discuss their applications, advantages, and limitations.
13. Discuss the significance of electrode material and design in resistance welding. How do electrode properties affect the efficiency and quality of the welding process?
14. Explain the effects of welding parameters, such as current, pressure, and time, on the quality of a resistance weld. How can these parameters be optimized for different materials?
15. Differentiate between soldering and brazing in terms of process, materials used, and applications. Discuss the factors that determine the selection of one process over the other in various engineering applications.

Short answer-type questions

1. What is the primary difference between MIG (metal inert gas) welding and TIG (tungsten inert gas) welding?
2. Explain the role of flux coating in shielded metal arc welding (SMAW).
3. What are the advantages of using DC over AC in arc welding?
4. Describe the function of a welding electrode in arc welding.
5. What is the purpose of preheating the base material before arc welding?
6. What are the two gases commonly used in oxyacetylene welding?
7. Explain the significance of the neutral flame in gas welding.
8. How does gas welding differ from arc welding in terms of the heat source?
9. What safety precautions should be taken when handling acetylene cylinders?
10. Why is flux often used in gas welding?

11. What is the basic principle behind resistance welding?
12. Differentiate between spot welding and seam welding.
13. How is the heat generated in resistance welding controlled?
14. What are the common applications of resistance welding in the automotive industry?
15. Explain the advantages of using resistance welding over other welding processes for joining thin sheets of metal.

Objective questions

1. Which of the following welding processes uses an electric arc to melt the workpieces? **[GATE 2018, ME]**
   (A) Gas welding   (B) Resistance welding   (C) Metal Arc Welding   (D) Brazing
2. In gas welding, the most commonly used fuel gas is: **[SSC JE 2019, ME]**
   (A) Propane   (B) Hydrogen   (C) Acetylene   (D) Methane
3. Which of the following welding processes is primarily used for joining thin sheets of metal?
   [IES 2017, ME]
   (A) Shielded Metal Arc Welding (SMAW)
   (B) Resistance Spot Welding
   (C) Submerged Arc Welding (SAW)
   (D) Gas Tungsten Arc Welding (GTAW)
4. In Metal Arc Welding, which of the following functions does the flux coating on the electrode perform? **[GATE 2020, ME]**
   (A) Provides filler material
   (B) Protects the weld pool from atmospheric contamination
   (C) Increases the welding speed
   (D) Reduces the welding temperature
5. In resistance welding, the heat generated at the weld joint is proportional to: **[IES 2018, ME]**
   (A) Voltage   (B) Current   (C) Square of the current   (D) Temperature difference
6. Which of the following processes involves joining metals by heating them to a temperature below their melting point and using a filler metal with a lower melting point? **[SSC JE 2020, ME]**
   (A) Brazing   (B) Soldering   (C) Welding   (D) Forge welding
7. In brazing, the filler metal typically has a melting point: **[GATE 2019, ME]**
   (A) Above 1,000 °C
   (B) Below 100 °C
   (C) Above 450 °C but below the melting point of the base metals
   (D) Above the melting point of the base metals
8. Which of the following is a characteristic advantage of Gas Tungsten Arc Welding (GTAW)? **[IES 2019, ME]**
   (A) High welding speed
   (B) Suitable for thick sections
   (C) High-quality and precise welds
   (D) Less expensive equipment
9. In soldering, the joint strength is typically: **[SSC JE 2018, ME]**
   (A) Higher than brazing
   (B) Lower than welding and brazing
   (C) The same as welding
   (D) Higher than welding but lower than brazing

10. The primary purpose of using flux in brazing is to: **[GATE 2021, ME]**
    (A) Reduce the melting point of the base metals.
    (B) Prevent oxidation and remove oxides from the surfaces to be joined.
    (C) Increase the strength of the joint.
    (D) Enhance the appearance of the joint.

Answers
1(C)   2(C)   3(B)   4(B)   5(C)   6(B)   7(C)   8(C)   9(B)   10(B)

Chapter 6
Advanced additive manufacturing

6.1 Concept of digital twin

In recent years, rapid advancements in the digital world and the seamless integration of information and communication technologies (ICT) with industrial operational technologies (OT) have significantly transformed the manufacturing industry. Innovations such as interconnected intelligent components on factory floors, the Internet of things (IoT), sensor data fusion, and cloud computing (CC) have paved the way for a new era known as smart manufacturing or digitized manufacturing.

Industry 4.0 is driving manufacturing toward autonomous and intelligent information exchange, machine control, and interoperable production systems. A key objective of Industry 4.0 is to enhance connectivity and integration within the production environment. To address existing challenges and further support this industrial revolution, the concept of "digital twin" has been introduced.

A digital twin is a virtual representation of a physical system, entity, or process. It integrates data from multiple sources to create a dynamic, real-time model that mirrors the behavior, conditions, and performance of its physical counterpart. This technology plays a crucial role in optimizing operations, predicting maintenance needs, and improving overall efficiency in smart manufacturing. The concept involves the following key components:

1. **Physical entity:**
   - The actual physical object, system, or process being modeled, such as a machine, a production line, or an entire manufacturing plant.
2. **Digital model:**
   - A virtual replica of the physical entity that simulates its physical and operational characteristics. This model is created using data from sensors, IoT devices, and other sources.
3. **Data integration:**
   - Real-time data is collected from the physical entity through sensors and monitoring systems. This data is continuously fed into the digital model to keep it updated and reflective of the current state of the physical entity.
4. **Simulation and analysis:**
   - The digital twin allows for the simulation, analysis, and optimization of the physical entity. This includes running "what-if" scenarios, predicting outcomes, and analyzing performance under various conditions.
5. **Feedback loop:**
   - Insights gained from the digital twin can be used to make adjustments or improvements to the physical entity, creating a feedback loop that enhances performance and efficiency.

https://doi.org/10.1515/9783112205891-006

6.1.1 Applications of digital twin in manufacturing

The use of digital twins in manufacturing is transforming the industry by enhancing efficiency, reducing costs, and improving decision-making. Key applications include

1. **Predictive maintenance:**
   – **Condition monitoring:** Digital twins can monitor the condition of machinery in real-time, detecting signs of wear, damage, or malfunctions before they lead to failure.
   – **Maintenance scheduling:** By analyzing data trends and predicting potential failures, digital twins help in scheduling maintenance activities proactively, reducing downtime and extending equipment lifespan.
2. **Production optimization:**
   – **Process simulation:** Digital twins enable manufacturers to simulate and optimize production processes, improving efficiency, throughput, and quality.
   – **Resource allocation:** By analyzing the performance of different production lines and equipment, digital twins help optimize resource allocation and balance workloads.
3. **Product development:**
   – **Design validation:** Digital twins allow for virtual testing and validation of product designs under various conditions before physical prototypes are built, reducing development time and costs.
   – **Customization:** Manufacturers can use digital twins to create customized products based on real-time data from customers, improving product fit and performance.
4. **Quality control:**
   – **Real-time monitoring:** Digital twins provide real-time visibility into production quality, enabling quick identification and correction of defects.
   – **Process improvement:** By analyzing quality data and production processes, digital twins help identify root causes of quality issues and implement corrective actions.
5. **Supply chain management:**
   – **Inventory management:** Digital twins can model inventory levels and predict supply chain disruptions, helping manufacturers manage stock levels and reduce inventory costs.
   – **Logistics optimization:** By simulating supply chain scenarios, digital twins assist in optimizing logistics and transportation strategies, improving delivery times and reducing costs.
6. **Energy management:**
   – **Energy consumption:** Digital twins help monitor and analyze energy consumption patterns, identify inefficiencies, and implement energy-saving measures.

- **Sustainability:** By optimizing energy usage and reducing waste, digital twins contribute to sustainability goals and reduce operational costs.
7. **Training and simulation:**
   - **Operator training:** Digital twins provide a realistic simulation environment for training operators and maintenance personnel, improving skills and reducing the risk of errors.
   - **Scenario testing:** Manufacturers can use digital twins to test different operational scenarios and responses, preparing for various contingencies and enhancing overall readiness.
8. **Asset management:**
   - **Lifecycle management:** Digital twins track the entire lifecycle of assets, from installation to decommissioning, providing insights into performance, maintenance needs, and replacement schedules.
   - **Cost analysis:** By analyzing data on asset usage and performance, digital twins help assess the total cost of ownership and make informed decisions about asset management.
9. **Customer experience:**
   - **Product usage insights:** Digital twins can provide insights into how customers use products, enabling manufacturers to offer better support, improve product designs, and tailor services to customer needs.
   - **Virtual showrooms:** Manufacturers can create virtual showrooms and product demonstrations using digital twins, enhancing the customer experience and facilitating sales.

6.1.2 Benefits of digital twin technology

- **Enhanced decision-making:** Provides real-time insights and predictive analytics, enabling better-informed decisions and faster response to issues.
- **Increased efficiency:** Optimizes production processes, reduces downtime, and improves resource allocation, leading to greater efficiency and productivity.
- **Cost savings:** Reduces costs associated with maintenance, inventory, and waste by enabling proactive management and optimization.
- **Improved quality:** Enhances product and process quality through real-time monitoring, analysis, and continuous improvement.
- **Innovation and agility:** Accelerate product development and innovation by enabling virtual testing and simulation, reducing time-to-market.

6.1.3 Key characteristics of digital twin technology

– **Real-time reflection:** The virtual space is highly synchronized with the physical environment, providing a real-time, multi-level representation of its counterpart.
– **Interaction and convergence**: This characteristic is categorized into three levels:
  i. **Interaction within the physical space:** A fully integrated system where different phases, components, and services interact seamlessly. Data generated at various stages are interconnected and easily accessible.
  ii. **Integration of historical and real-time data:** A comprehensive digital twin incorporates both multi-physics models and data-driven approaches, combining domain expertise with real-time operational insights.
  iii. **Connection between physical and virtual spaces:** The digital twin serves as a unified platform, ensuring smooth, bidirectional communication between the real and digital environments.
– **Self-evolution**: A digital twin dynamically updates its data in real time, ensuring an accurate reflection of the physical system. Its parallel connectivity facilitates continuous comparison between the virtual and physical spaces, enabling ongoing optimization and enhancement of the digital model.

6.1.4 Challenges and considerations

– **Data integration:** Effective implementation requires seamless integration of data from various sources, which can be complex, and requires robust data management systems.
– **Cost of implementation:** Initial setup costs, including sensors, software, and infrastructure, can be high, although they may be offset by long-term savings and benefits.
– **Data security:** Ensuring the security and privacy of data collected and used by digital twins is crucial to prevent unauthorized access and cyber threats.
– **Complexity:** Developing and maintaining accurate digital twins for complex systems or processes can be challenging and require specialized expertise.

6.1.5 Key applications of digital twins in product development

1. **Design validation and optimization**
   – **Virtual prototyping:** Digital twins enable the creation of virtual prototypes, allowing designers to test and validate product designs in a simulated environment. This helps identify potential issues early in the development process and reduce the need for physical prototypes.

– **Design iteration:** Engineers can run simulations on digital twins to explore various design alternatives, optimize performance, and make data-driven decisions without incurring the costs and time associated with physical testing.

2. **Performance simulation**
   – **Stress testing:** Digital twins allow for the simulation of how a product will perform under different stress conditions, such as extreme temperatures, loads, or environmental factors. This helps ensure that the product meets safety and reliability standards.
   – **Behavior prediction:** Engineers can predict how the product will behave in real-world conditions, including wear and tear, fatigue, and failure modes, by analyzing the digital twin's response to simulated scenarios.

3. **Integration and interoperability testing**
   – **System integration:** Digital twins enable the testing of how different components of a product interact with each other and with other systems. This is particularly useful for complex products with multiple subsystems or for products that need to integrate with existing infrastructure.
   – **Interoperability:** By simulating interactions between various components and systems, digital twins help identify compatibility issues and ensure that the product functions seamlessly within its intended ecosystem.

4. **Customization and personalization**
   – **Customer requirements:** Digital twins allow manufacturers to model and test customizations based on individual customer requirements, such as tailored features or specifications. This ensures that personalized products meet customer expectations and performance standards.
   – **Configurable designs:** Engineers can use digital twins to explore different configuration options and their impacts, allowing for the development of products that can be easily adapted to various use cases or preferences.

5. **Cost reduction**
   – **Reduced prototyping costs:** By enabling virtual testing and validation, digital twins reduce the need for physical prototypes, thereby saving costs associated with materials, manufacturing, and testing.
   – **Efficient development:** The ability to quickly iterate on designs and make data-driven decisions accelerates the development process, reducing time-to-market and associated costs.

6. **Enhanced collaboration**
   – **Cross-functional teams:** Digital twins facilitate collaboration among cross-functional teams by providing a shared, up-to-date model of the product. This helps synchronize efforts across design, engineering, manufacturing, and other departments.
   – **Remote access:** Teams can access and work with digital twins remotely, enabling collaboration with global stakeholders and reducing the need for physical meetings or site visits.

7. **Lifecycle management**
   - **Continuous improvement:** Digital twins provide insights throughout the product lifecycle, from design through operation and maintenance. This ongoing feedback allows for continuous improvement of the product and informs future development efforts.
   - **End-of-life planning:** Engineers can use digital twins to analyze the end-of-life phase of a product, including decommissioning or recycling processes, to design for sustainability and minimize environmental impact.

8. **User experience enhancement**
   - **Simulation of usage scenarios:** Digital twins enable the simulation of various usage scenarios to assess how different user behaviors affect product performance. This helps in designing products that offer an optimal user experience.
   - **User feedback integration:** By analyzing data from the digital twin, manufacturers can incorporate user feedback into the product design, ensuring that the final product meets customer needs and preferences.

6.1.6 Benefits of using digital twins in product development

- **Accelerated time-to-market:** By streamlining design, testing, and optimization processes, digital twins help speed up the development cycle and bring products to market more quickly.
- **Improved product quality:** Real-time simulations and validations lead to higher-quality products with fewer defects and better performance.
- **Cost savings:** Reduced need for physical prototypes and faster development cycles contribute to lower overall development costs.
- **Enhanced innovation:** The ability to explore and test various design options in a virtual environment fosters innovation and creativity in product development.
- **Better decision-making:** Data-driven insights from digital twins support informed decision-making throughout the development process, leading to more effective and strategic choices.

Summary: Digital twin technology represents a significant advancement in manufacturing by creating virtual replicas of physical entities that integrate real-time data for simulation, analysis, and optimization. Its applications span predictive maintenance, production optimization, product development, quality control, supply chain management, energy management, training, asset management, and customer experience. While offering substantial benefits such as enhanced decision-making, increased efficiency, and cost savings, digital twin technology also presents challenges related to data integration, implementation costs, security, and complexity.

6.2 Case studies on digital twin implementations in industry

Here are some notable case studies showcasing how various industries have successfully implemented digital twins to drive innovation, efficiency, and performance improvements:

1. General Electric (GE): digital twin in aviation
– **Industry:** Aviation
– **Implementation:** General Electric (GE) implemented digital twin technology for its aircraft engines. By creating digital replicas of physical engines, GE could monitor the engines' performance in real-time and predict maintenance needs.
– **Results:**
  – **Predictive maintenance:** Digital twins enabled predictive maintenance, reducing unplanned downtime by up to 25%.
  – **Fuel efficiency:** GE improved fuel efficiency by optimizing engine performance based on real-time data from digital twins.
  – **Cost savings:** Using GE's engines, airlines saw significant cost savings through reduced maintenance and improved fuel efficiency.

2. Siemens: digital twin in manufacturing
– **Industry:** Manufacturing
– **Implementation:** Siemens implemented digital twins in its Amberg Electronics Plant in Germany. The plant produces programmable logic controllers (PLCs), and digital twins were used to simulate and optimize production processes.
– **Results:**
  – **Increased productivity:** Siemens achieved a 75% increase in productivity by using digital twins to optimize production workflows.
  – **Quality improvement:** The digital twin technology has contributed to a near-zero defect rate in the manufacturing process.
  – **Energy efficiency:** Siemens also improved energy efficiency in the plant by 20%, aligning it with its sustainability goals.

3. Unilever: digital twin in consumer goods manufacturing
– **Industry:** Consumer goods
– **Implementation:** Unilever applied digital twin technology at its factory in Valinhos, Brazil, which produces various personal care products. The digital twin modeled the entire production line to optimize operations and reduce waste.
– **Results:**
  – **Waste reduction:** The digital twin helped Unilever reduce waste by 50% by optimizing production parameters and minimizing material usage.
  – **Increased throughput:** Production throughput increased by 15% due to the optimization of processes identified through the digital twin.

- **Sustainability:** Unilever has improved its sustainability metrics by reducing energy consumption and minimizing the environmental impact of its production processes.

4. Rolls-Royce: digital twin in marine industry
- **Industry:** Marine
- **Implementation:** Rolls-Royce developed digital twins for its marine engines and ship systems. These digital twins provide real-time monitoring and analytics to enhance the performance and reliability of maritime operations.
- **Results:**
  - **Operational efficiency:** Ships equipped with Rolls-Royce digital twins experienced up to a 20% increase in operational efficiency by optimizing fuel usage and engine performance.
  - **Predictive maintenance:** The digital twins enable predictive maintenance, reducing engine-related failures and associated costs.
  - **Reduced emissions:** Optimized engine performance has led to a significant reduction in CO_2 emissions, contributing to environmental sustainability.

5. Daimler AG: digital twin in automotive manufacturing
- **Industry:** Automotive
- **Implementation:** Daimler AG, the parent company of Mercedes-Benz, implemented digital twins in its car manufacturing plants. Digital twins are used to simulate and optimize the assembly line, product design, and production processes.
- **Results:**
  - **Production flexibility:** The digital twins allow Daimler to increase production flexibility, enabling the plant to adapt quickly to changes in demand and new product introductions.
  - **Quality assurance:** By simulating production processes, Daimler has improved the quality of its vehicles and reduced the number of defects.
  - **Cost reduction:** Digital twins contributed to a 10% reduction in production costs by streamlining processes and reducing material waste.

6. Philips Healthcare: digital twin in medical equipment
- **Industry:** Healthcare
- **Implementation:** Philips used digital twins to model and simulate the operation of its medical imaging equipment, such as MRI machines and CT scanners. The digital twins were used to monitor equipment performance and predict maintenance needs.
- **Results:**
  - **Improved uptime:** Digital twins help reduce equipment downtime by predicting when maintenance is needed, ensuring that machines were available when required.

- **Enhanced performance:** Philips has optimized the performance of its imaging equipment, resulting in better image quality and reduced scan times for patients.
- **Patient safety:** By using digital twins to monitor and control equipment, Philips enhanced patient safety by ensuring that the machines operated within safe parameters.

7. Shell: digital twin in oil and gas
- **Industry:** Oil and gas
- **Implementation:** Shell implemented digital twin technology for its offshore oil platforms. The digital twins provide a real-time virtual model of the platforms, enabling remote monitoring, simulation, and optimization of operations.
- **Results:**
  - **Operational safety:** The digital twins enhance safety by allowing Shell to simulate and plan complex operations, such as equipment shutdowns and startups, thereby reducing the risk of accidents.
  - **Cost efficiency:** Shell achieved significant cost savings by optimizing the use of resources, reducing the need for on-site inspections, and minimizing downtime.
  - **Sustainability:** The technology also helps Shell to reduce emissions by optimizing energy use on the platforms.

---

i | **Summary:** These case studies demonstrate the wide-ranging applications and benefits of digital twin technology across various industries. Digital twins have proven to be effective in enhancing operational efficiency, reducing costs, improving product quality, and contributing to sustainability goals. As digital twin technology continues to evolve, its adoption is expected to grow, offering even more innovative solutions for industrial challenges.

6.3 Overview of additive manufacturing technologies

Additive manufacturing (AM), commonly known as material printing, is a process of creating objects by adding material layer by layer, following a digital model. This method contrasts with subtractive manufacturing, where material is removed from a solid block to shape the final product.

AM enables the production of intricate geometries and customized designs that are often unachievable with traditional manufacturing techniques. It encompasses various technologies, each with distinct advantages and applications. From rapid prototyping and personalized manufacturing to fabricating complex, high-performance components for industries such as aerospace and healthcare, AM is transforming product development.

As advancements in AM continue, its applications are expected to grow, driving innovation, efficiency, and cost-effectiveness across multiple sectors.

6.3.1 Types of additive manufacturing technologies

1. Stereolithography (SLA)

– **Overview:** Stereolithography (SLA) is an AM technology that uses ultraviolet (UV) light to cure and solidify photopolymer resin. SLA machines are known for their high precision and excellent surface finish.
– **Process:** SLA uses a laser to cure and solidify layers of photopolymer resin in a vat. The laser traces each layer based on a digital model, and the platform lowers, allowing the next layer to be cured on top. Figure 6.1 shows the components of SLA.
– **Resin vat:** A vat filled with liquid photopolymer resin.
– **UV laser:** A UV laser or projector cures the resin layer by layer according to the digital design.
– **Layer-by-layer curing:** The build platform moves up or down to accommodate the addition of each layer of cured resin.
– **Post-processing:** After printing, the part is often washed to remove excess resin and further cured to achieve final hardness.

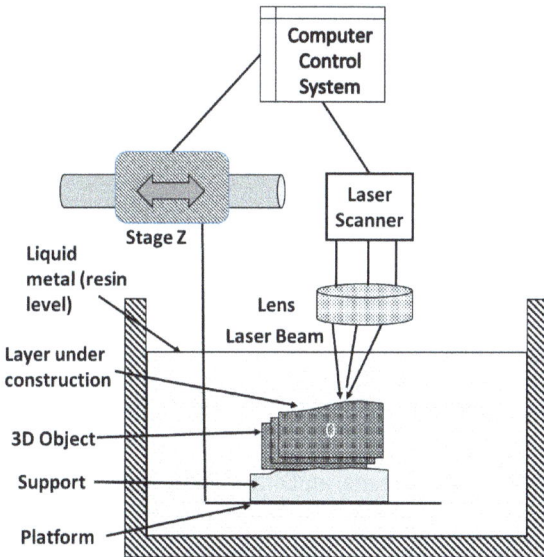

Figure 6.1: Components of SLA.

- **Materials:** Photopolymer resins.
- **Applications:**
  - **Prototyping:** Ideal for creating highly detailed prototypes with smooth surface finishes.
  - **Dental and medical models:** Used for producing accurate dental aligners, implants, and anatomical models.
  - **Jewelry:** Enables the creation of intricate and detailed jewelry designs.
- **Advantages:**
  - **High precision:** Excellent for fine details and smooth surface finishes.
  - **Versatility:** Different resins for various applications, including flexible and high-temperature materials.
- **Limitations:**
  - **Cost:** Resin and SLA machines can be more expensive compared to FDM.
  - **Post-processing:** Requires additional steps such as washing and curing.

2. Fused deposition modeling (FDM)

- **Process:** FDM works by extruding thermoplastic filaments through a heated nozzle, which deposits the material layer by layer to create the object. As each layer cools, it bonds to the one beneath it. The structure of FDM is illustrated in Figure 6.2.

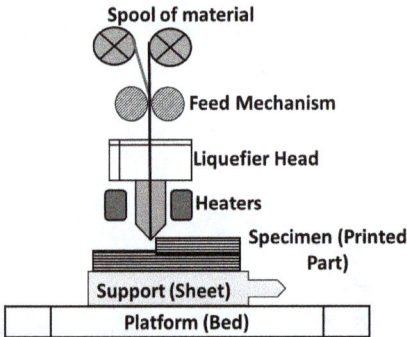

Figure 6.2: Components of FDM.

- **Materials:** ABS, PLA, PETG, nylon, polycarbonate, and other thermoplastics.
- **Applications:**
  - **Prototyping:** Widely used for creating functional prototypes and low-cost models.
  - **End-use parts:** Used for producing custom or low-volume end-use parts, such as brackets, housings, and fixtures.
  - **Educational tools:** Popular in schools and universities for teaching design and engineering concepts.

3. Selective laser sintering (SLS)

– **Overview:** Selective laser sintering (SLS) is an AM technology that utilizes a laser to sinter powdered materials, bonding them layer by layer to create a solid structure. It is particularly effective for producing complex and functional parts.
– **Process:** SLS employs a laser to sinter powdered materials, such as nylon or other polymers, layer by layer. The laser fuses the powder particles to form a solid structure, with the process continuing until the complete object is created. The components of SLS are illustrated in Figure 6.3.
  – **Powder bed:** A build chamber is filled with powdered material.
  – **Laser sintering:** A laser selectively heats and fuses the powder according to the digital design, layer by layer.
  – **Layer-by-layer fusion:** The build platform lowers, and a new layer of powder is spread over the previous layer, which is then sintered.
  – **Cooling:** Once printing is complete, the part is allowed to cool before being removed from the powder bed.

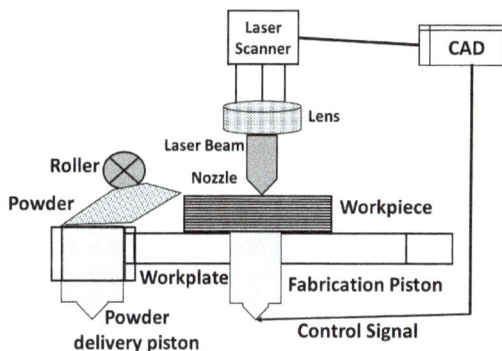

Figure 6.3: Components of SLS.

– **Materials:** Nylon, polyamide, thermoplastic elastomers, metal powders.
– **Applications:**
  – **Functional prototypes:** Ideal for creating durable prototypes with complex geometries and moving parts.
  – **End-use parts:** Used in industries such as aerospace, automotive, and medical for low-volume production of durable parts.
  – **Custom manufacturing:** Enables the production of customized parts, such as orthopedic implants and prosthetics.
– **Advantages:**
  – **Complex geometries:** Capable of producing intricate and complex shapes that are difficult to achieve with traditional methods.
  – **Material efficiency:** The unused powder supports the part during printing, reducing material waste.

- **Limitations:**
    - **Cost:** SLS machines and materials can be relatively expensive.
    - **Surface finish:** Parts may require additional post-processing to improve surface finish and remove powder residue.

4. Digital light processing (DLP)

- **Process:** Similar to SLA, DLP uses a digital light projector to flash entire layers of UV light onto a vat of photopolymer resin, curing the resin layer by layer. (See Figure 6.4)

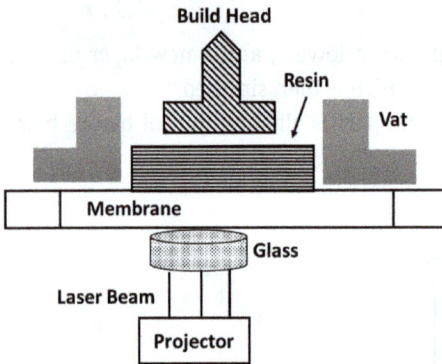

Figure 6.4: Components of DLP.

- **Materials:** Photopolymer resins.
- **Applications:**
    - **High-resolution prototyping:** Produces detailed parts with fine features and smooth finishes.
    - **Dental and medical models:** Used for creating dental molds, surgical guides, and other medical applications.
    - **Jewelry:** Ideal for intricate jewelry designs with high detail.

5. Binder jetting

- **Process:** In binder jetting, a liquid binder is selectively applied to a powder bed, bonding the particles together. This process is repeated layer by layer until the object is fully formed. The printed part is then usually sintered or infused with metal to attain its final properties (see Figure 6.5).
- **Materials:** Sand, ceramics, metals, and composites.
- **Applications:**
    - **Sand casting molds:** Used to create sand molds and cores for metal casting.
    - **Metal parts:** Produces metal components that can be sintered or infiltrated to achieve the desired mechanical properties.
    - **Architectural models:** Used for creating large-scale models of buildings and other structures.

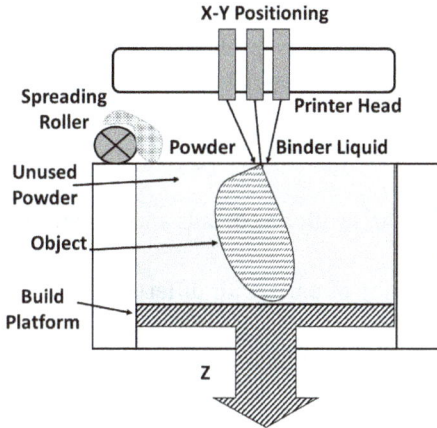

Figure 6.5: Components of binder jetting.

6. Direct metal laser sintering (DMLS) and selective laser melting (SLM)

- **Process:** Both DMLS and SLM use a laser to sinter or melt metal powder layer by layer. DMLS partially melts the powder, while SLM fully melts it, resulting in solid metal parts.
- **Materials:** Stainless steel, aluminum, titanium, cobalt-chrome, and other metal alloys.
- **Applications:**
  - **Aerospace components:** Used for producing complex, lightweight metal parts for aerospace applications.
  - **Medical implants:** Ideal for manufacturing custom implants, such as hip and knee replacements.
  - **Tooling:** Used for creating durable, high-precision tools for injection-molding and other manufacturing processes.

7. Electron beam melting (EBM)

- **Process:** EBM uses an electron beam to melt metal powder in a vacuum environment. The process is similar to SLM but operates at higher temperatures and utilizes different materials.
- **Materials:** Titanium, Inconel, cobalt-chrome, and other high-performance alloys.
- **Applications:**
  - **Aerospace and defense:** Used for manufacturing high-strength, lightweight parts that can withstand extreme conditions.
  - **Medical implants:** Produces biocompatible implants with complex geometries.
  - **Turbine blades:** Suitable for producing turbine blades and other high-temperature components.

8. Material jetting

- **Process:** Material jetting involves depositing droplets of photopolymer or wax material layer by layer using inkjet printheads. The material is cured or solidified as each layer is deposited.
- **Materials:** Photopolymers, waxes, and other jettable materials.
- **Applications:**
  - **High-resolution prototypes:** Produce parts with fine details and smooth finishes, ideal for visual prototypes.
  - **Multi-material printing:** Enables printing of parts with different materials and colors in a single build.
  - **Dental and medical models:** Used for creating highly accurate dental and anatomical models.

9. Sheet lamination (laminated object manufacturing – LOM)

- **Process:** In sheet lamination, thin layers of material, such as paper, plastic, or metal, are bonded together using heat and pressure. The layers are then cut into shape using a laser or blade.
- **Materials:** Paper, plastic, metal foil, and composite sheets.
- **Applications:**
  - **Prototyping:** Used for creating large-scale prototypes and models.
  - **Tooling:** Suitable for producing molds and tooling components.
  - **Art and design:** Popular for artistic and architectural models due to its ability to produce detailed, layered structures.

10. Vat photopolymerization (continuous liquid interface production – CLIP)

- **Process:** CLIP is a continuous version of SLA, where a continuous liquid interface of uncured resin is maintained, allowing for rapid, layerless printing. The part is drawn out of the resin as it solidifies.
- **Materials:** Photopolymer resins.
- **Applications:**
  - **Rapid prototyping:** Enables fast production of prototypes with smooth surfaces.
  - **Medical devices:** Used for creating intricate medical devices and implants.
  - **Automotive components:** Suitable for producing lightweight, detailed components for automotive applications.

Summary: FDM, SLA, and SLS are three prominent AM technologies, each with its own strengths and applications:
- **FDM** is known for its cost-effectiveness and ease of use, making it ideal for prototyping and educational purposes.

- **SLA** offers high precision and excellent surface finish, making it suitable for detailed prototypes and specialized applications.
- **SLS** excels in producing complex geometries and functional parts, making it valuable for both prototyping and low-volume production.

Each technology has its unique advantages and limitations, and the choice of technology depends on the specific requirements of the application.

6.3.2 Applications of additive manufacturing technologies

1. **Aerospace:**
   - **Lightweight components:** AM is used to produce complex, lightweight parts that reduce fuel consumption and enhance performance in aerospace applications.
   - **Rapid prototyping:** Enables the quick development and testing of aerospace components, reducing the time and cost of bringing new designs to market.
2. **Automotive:**
   - **Custom parts:** AM allows for the production of custom and replacement parts on demand, reducing inventory and lead times.
   - **Tooling:** Used to create high-precision, durable tools for manufacturing processes, improving efficiency and reducing costs.
3. **Healthcare:**
   - **Medical implants:** AM is used to produce patient-specific implants, such as dental implants, hip and knee replacements, and cranial implants, thereby improving patient outcomes.
   - **Surgical guides:** Enable the creation of custom surgical guides that assist surgeons in performing precise procedures.
4. **Consumer goods:**
   - **Personalized products:** AM allows for the mass customization of consumer products, such as footwear, eyewear, and jewelry, catering to individual preferences.
   - **Prototyping:** Used for rapidly developing and iterating on consumer product designs, thereby reducing time-to-market.
5. **Industrial manufacturing:**
   - **Spare parts:** AM enables the on-demand production of spare parts, reducing downtime and inventory costs for manufacturers.
   - **Tooling and jigs:** Used to produce custom tooling, jigs, and fixtures that improve manufacturing efficiency and quality.

6. **Architecture and construction:**
   - **Architectural models:** AM is used to create detailed architectural models that help visualize complex designs and improve communication with clients.
   - **Construction components:** AM technologies, such as large-scale 3D printing, are being explored for producing building components and even entire structures.
7. **Art and design:**
   - **Sculptures and artwork:** AM allows artists and designers to create intricate sculptures and artwork that would be difficult or impossible to produce with traditional methods.
   - **Custom jewelry:** Enables the creation of highly detailed and personalized jewelry designs.

6.3.3 Additive manufacturing: materials and their properties

Additive manufacturing (AM) technologies utilize a variety of materials, each with distinct properties tailored to specific applications. Here is an overview of common AM materials and their properties:

1. Thermoplastics: Thermoplastics are widely used in AM due to their versatility, ease of processing, and availability in various grades.
- **Acrylonitrile butadiene styrene (ABS):**
  - Properties: Good mechanical strength, impact resistance, and toughness. It has a high melting point and is relatively easy to print.
  - Applications: Prototyping, functional parts, consumer goods, and automotive components.
- **Polylactic acid (PLA):**
  - Properties: Biodegradable, low melting point, and ease of use. Offers good surface finish and dimensional accuracy but has lower mechanical strength compared to ABS.
  - Applications: Prototyping, educational models, and non-functional decorative items.
- **Polyethylene terephthalate glycol (PETG):**
  - Properties: Good strength, flexibility, and impact resistance. It is also moisture-resistant and has a relatively low warping tendency.
  - Applications: Functional prototypes, mechanical parts, and containers.
- **Nylon (polyamide):**
  - Properties: High strength, durability, and flexibility. It exhibits good abrasion resistance and impact resistance.
  - Applications: Functional parts, gears, and wear-resistant components.

- **Polycarbonate (PC):**
  - Properties: High impact resistance, strength, and heat resistance. It has a high glass transition temperature.
  - Applications: Engineering prototypes, durable parts, and high-strength components.

2. Photopolymers: Photopolymers are used in resin-based AM technologies like SLA and DLP.
- **Standard photopolymer resins:**
  - Properties: Typically they have good detail resolution and a smooth surface finish. However, they are less durable and more brittle compared to thermoplastics.
  - Applications: Detailed prototypes, dental models, and jewelry.
- **Tough resins:**
  - Properties: Enhanced impact resistance and mechanical strength compared to standard resins. The aim is to combine the precision of photopolymers with greater durability.
  - Applications: Functional prototypes and end-use parts requiring better durability.
- **Flexible resins:**
  - Properties: Rubber-like flexibility and elongation properties. They offer good shock absorption and can simulate various elastomeric materials.
  - Applications: Flexible parts, custom grips, and wearables.
- **High-temperature resins:**
  - Properties: Capable of withstanding high temperatures without significant deformation. They are often used in high-heat applications.
  - Applications: Parts for high-temperature environments and tooling.

3. Metals: Metal AM materials are used in technologies like DMLS, SLM, and EBM.
- **Stainless steel:**
  - Properties: High strength, corrosion resistance, and good mechanical properties. It is suitable for a wide range of applications.
  - Applications: Functional parts, aerospace components, and medical implants.
- **Titanium:**
  - Properties: High strength-to-weight ratio, corrosion resistance, and biocompatibility. It is also lightweight and durable.
  - Applications: Aerospace parts, medical implants, and high-performance engineering components.
- **Aluminum:**
  - Properties: Lightweight, good thermal and electrical conductivity, and moderate strength. It is easy to process and machine.
  - Applications: Lightweight parts, automotive components, and heat exchangers.

- **Cobalt-chrome:**
  - Properties: High hardness, wear resistance, and biocompatibility. It has excellent corrosion resistance and is used in high-stress applications.
  - Applications: Medical implants, aerospace components, and high-wear parts.

4. Ceramics: Ceramic materials are used for high-temperature applications and require special AM techniques.
- **Alumina (aluminum oxide):**
  - Properties: High hardness, wear resistance, and thermal stability. It is also chemically inert.
  - Applications: Cutting tools, wear-resistant parts, and high-temperature applications.
- **Zirconia:**
  - Properties: High toughness, strength, and wear resistance. It also has good thermal insulation properties.
  - Applications: Dental crowns, structural ceramics, and wear-resistant components.

5. Composites: Composite materials are engineered by combining a matrix material with reinforcing fibers or particles to enhance mechanical, thermal, and chemical properties. The matrix, typically made of polymers, metals, or ceramics, serves as a binding agent, while the reinforcement – such as carbon fibers, glass fibers, or ceramic particles – provides strength, stiffness, and durability. This combination results in materials that are lightweight yet strong, corrosion-resistant, and capable of withstanding high stress and environmental challenges. Composites are widely used in industries such as aerospace, automotive, construction, and sports equipment, where performance and weight reduction are critical. The ability to customize their properties by varying the type, orientation, and volume of reinforcement makes composite materials highly versatile and suitable for a broad range of applications.
- **Carbon fiber-reinforced polymers (CFRP):**
  - **Properties:** High strength-to-weight ratio, stiffness, and dimensional stability. They are also lightweight and durable.
  - **Applications:** Aerospace components, automotive parts, and high-performance sports equipment.
- **Glass fiber-reinforced polymers (GFRP):**
  - **Properties:** Good strength, impact resistance, and cost-effectiveness compared to CFRP. It also has moderate weight.
  - **Applications:** Automotive parts, consumer goods, and construction materials.

6.3.4 Emerging trend in additive manufacturing

Additive manufacturing (AM) is rapidly evolving, with several emerging trends re-shaping its applications and potential. The most commonly known method is 3D print-ing. Here are some key trends in the field:

1. **Advanced materials development**
   - **High-performance polymers:** Development of materials such as PEEK, PEKK, and ULTEM for aerospace and medical applications.
   - **Metal alloys:** New alloys such as titanium aluminides, high-entropy alloys, and lightweight aluminum-based alloys.
   - **Bio-inks:** Biocompatible materials for bioprinting tissues and organs.
   - **Sustainable materials:** Use of recycled and biodegradable materials for en-vironmentally friendly production.

2. **Hybrid manufacturing**
   - Integration of additive and subtractive manufacturing (e.g., CNC machining and AM) for enhanced precision and finishing.
   - Combining AM with traditional manufacturing methods to optimize produc-tion workflows.

3. **4D Printing**
   - **Smart materials:** Development of materials that can change shape, proper-ties, or function over time in response to external stimuli such as heat, light, or moisture.
   - Applications in aerospace, robotics, and medical devices.

4. **Large-scale additive manufacturing (LSAM)**
   - Techniques like wire-arc AM (WAAM) and robotic 3D printing enable the pro-duction of large structures such as aerospace components, buildings, and ma-rine vessels.

5. **AI and machine learning integration**
   - AI-driven design optimization and process control for improved efficiency, reduced defects, and predictive maintenance.
   - Generative design to create lightweight, structurally optimized components.

6. **Advancements in speed and precision**
   - **Multi-laser systems:** Multiple lasers in metal powder bed fusion systems for faster production.
   - **Continuous liquid interface production (CLIP):** Faster and more precise printing using liquid resin.

7. **Medical applications and personalization**
   - Growth in custom prosthetics, implants, and surgical tools tailored to individ-ual patients.
   - Bioprinting of tissues and organs for regenerative medicine and drug testing.

8. **AM for aerospace and defense**
   - Increased adoption due to weight reduction, part consolidation, and on-demand spare part manufacturing.
   - Enhanced materials designed to withstand extreme temperature and stress requirements.
9. **On-demand and distributed manufacturing**
   - Decentralized production with 3D printers located closer to end-users or in remote locations for spare parts reduces supply chain dependencies.
10. **Sustainability focus**
    - AM is increasingly viewed as a sustainable technology due to reduced material waste and energy consumption.
    - Recycling of materials used in AM processes.
11. **Post-processing innovations**
    - Automation in post-processing for cleaning, surface finishing, and heat treatment.
    - Enhanced post-processing techniques to meet stringent quality standards for industries such as aerospace and healthcare.
12. **Standardization and certification**
    - Development of international standards for AM materials, processes, and products to ensure quality and reliability.
    - Certification pathways for critical components, especially in regulated industries.

6.4 Understanding 3D printing

3D printing, also called additive manufacturing (AM), is a technique that fabricates three-dimensional objects by progressively layering material according to a digital blueprint. Unlike conventional manufacturing, which typically involves cutting or removing material from a solid piece, this method builds structures from the ground up, minimizing waste and enabling intricate designs. The basic steps in operating a 3D printer has been elaborated in figure 6.6.

Key aspects of 3D printing
1. **Digital model:** A 3D model is developed using computer-aided design (CAD) software and then divided into thin slices, which guide the printer's movements.
2. **Printing process:** Material is systematically deposited, fused, or solidified layer by layer to form the final object. Popular techniques include FDM, SLA, and SLS.
3. **Post-processing:** Once printing is complete, additional steps such as removing supports, refining surfaces, or curing may be required to enhance the final product.

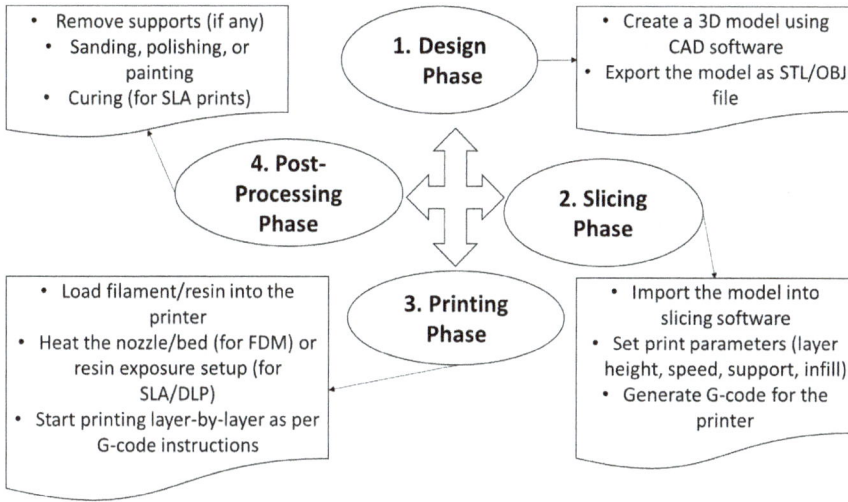

Figure 6.6: steps in operating 3D printer.

6.4.1 Application of tolerances and fitments considering 3D printing processes

Tolerances and **fitments** are crucial in ensuring that 3D-printed parts meet functional requirements and fit correctly within assemblies. Due to the nature of 3D printing, certain considerations must be made:

1. Understanding tolerances

– **Tolerance definition:** Tolerance refers to the allowable deviation from a specified dimension. In 3D printing, it impacts how precisely the printer can reproduce dimensions.
– **Impact of technology:**
  – **FDM:** Typically has larger tolerances due to layer-by-layer deposition and the potential for warping. Tolerances can vary based on printer quality and the material used.
  – **SLA:** Generally offers higher precision and finer tolerances due to the nature of photopolymer curing, resulting in smoother surfaces and more accurate dimensions.
  – **SLS:** Provides good dimensional accuracy and can handle complex geometries, though it may exihibit slight variations due to powder spreading and sintering.
– **Design considerations:**
  – **Over-compensation:** Design features may need to be slightly adjusted to account for material shrinkage or expansion during printing and post-processing.

– **Clearances:** Provide appropriate clearances between mating parts to ensure proper fit and functionality, especially considering potential post-printing changes such as warping.

2. Fitments in 3D printing

– **Fitment definition:** Fitment refers to how well parts fit together in an assembly, including tolerances for parts that need to be assembled or interfaced with one another.
– **Types of fits:**
   – **Loose fit:** Where parts are designed with extra clearance for ease of assembly, often used in cases where parts need to move or be adjusted.
   – **Press fit:** Where parts are designed to fit tightly together, requiring precise tolerances to ensure a secure fit without gaps.
   – **Interference fit:** Where parts are intentionally designed with a slight interference to create a strong, tight fit, often requiring post-processing or additional force to assemble.
– **Design for fitment:**
   – **Assembly considerations:** Design parts with assembly in mind, ensuring that tolerances are accounted for to avoid issues such as binding or excessive play.
   – **Adjustments:** Utilize design software to simulate fitments and tolerances, adjusting dimensions as needed based on the chosen 3D-printing technology and material properties.

3. Factors affecting tolerances and fitments

– **Material properties:** Different materials have different shrinkage and expansion characteristics, affecting final dimensions and fit. For example, thermoplastics used in FDM may shrink more than resins used in SLA.
– **Printer calibration:** The accuracy of the 3D printer, including its calibration and maintenance, can impact the final tolerances and fitments.
– **Post-processing effects:** Additional processing steps like sanding, polishing, or curing can alter dimensions and affect fit.

Summary: Understanding and managing tolerances and fitments are crucial in 3D printing to ensure that parts function correctly and fit together as intended. While 3D-printing technologies offer flexibility and customization, designers must account for the specific characteristics of each technology and material to achieve the desired results. By considering these factors during the design phase and utilizing appropriate post-processing techniques, the precision and quality of 3D-printed parts can be optimized for their intended applications.

6.4.2 Understanding process algorithm of slicing software and slicing techniques

1. Slicing software: Slicing software (also known as slicers) converts 3D models into instructions that 3D printers can understand, typically in the form of G-code. This process involves several key steps:

i) Model input
The process begins when a user imports a 3D model, usually in STL, OBJ, or 3MF format, into the slicer. The software reads the geometry, which is typically represented as a mesh of triangles, and prepares it for slicing.

ii) Orientation and layout
Users can orient the model, scale it, or position it on the virtual print bed. Proper orientation can reduce print time, improve surface quality, and minimize the need for support structures.

iii) Layer slicing
The core function of slicing software is to break the 3D model into horizontal layers. This step divides the model into thin 2D cross-sections, which the printer will later build up layer by layer. The thickness of these layers (layer height) is determined by the user and influences both print quality and speed.

iv) Path planning (toolpath generation)
Once the model is sliced into layers, the slicer generates toolpaths for the printer. This is the path that the print head will follow to extrude material:
– **Infill pattern generation**: The slicer determines how the internal structure of each layer is filled. Various patterns (grid, honeycomb, triangle, etc.) are used to balance strength, weight, and material usage.
– **Perimeter and shells**: The outermost walls (perimeters) of the object are defined to ensure a smooth surface finish. Multiple shells (walls) can be added to enhance strength.
– **Support structure generation**: If there are overhanging sections that can't be printed in midair, support structures are generated to provide temporary material that can be removed later.

v) Parameter assignment
The slicer allows the user to specify various parameters for printing:
– **Print speed**: Affects the time taken to complete the print and the print quality.
– **Extrusion temperature**: Determines the temperature at which the filament is melted and extruded.
– **Bed temperature**: Ensures proper adhesion of the print to the bed.

vi) G-code generation

The final output of the slicing process is a G-code file. This file contains machine-specific instructions, including the position of the print head, extrusion rates, speed, and fan control, instructing the 3D printer on how to recreate the 3D model layer by layer.

2. Slicing techniques

Slicing techniques vary depending on the material being printed, the type of 3D printer, and the required print quality. Some common slicing techniques include:

i) Layer height adjustment (variable layer height)

Different parts of the print can benefit from varying layer heights:

- **Thicker layers (0.2–0.3 mm)**: Used for faster printing and stronger parts but with less detail.
- **Thinner layers (0.1 mm or lower)**: Used for detailed, high-quality surfaces but slower to print.

Some slicers allow variable layer height, which automatically adjusts the height in areas that require more detail.

ii) Shell and infill control

- **Increasing shells**: Adding more outer layers or shells improves strength and durability. This is useful for parts that need to be strong on the outside.
- **Infill density**: Lower infill percentages (5–20%) reduce material usage and weight, whereas higher infill percentages (50–100%) are used for stronger, more solid objects.

iii) Support structure optimization

- **Tree supports**: Curved or branched supports that are more efficient in providing support to overhangs without excessive material waste.
- **Grid supports**: Simple grid-like structures that are easy to remove but can be difficult to print without proper adhesion.

Some slicers offer support placement only in specific areas to reduce material usage and simplify post-processing.

iv) Bridging and overhangs

Slicers often include algorithms to manage how filament is printed over gaps (bridging) or steep angles (overhangs). This ensures that the filament doesn't droop or sag when printed without support.

v) Speed and acceleration control
To avoid print defects, slicers allow control over the speed of different operations:
- **Slower speeds for small parts and intricate details** to ensure quality.
- **Faster speeds for infill** to reduce overall print time without sacrificing external quality.

vi) Retraction settings
Retraction prevents stringing by pulling the filament back into the nozzle when the print head moves between different areas of the print. Tuning retraction distance and speed helps prevent oozing and results in cleaner prints.

vii) Adaptive slicing
Adaptive slicing changes the layer height dynamically based on the curvature or complexity of the model. This technique allows for faster prints without compromising detail in areas where high resolution is required.

- **Conclusion:** Slicing software is essential for preparing 3D models for printing, with algorithms that optimize the layer-by-layer building process. Through various slicing techniques, users can control print quality, speed, material usage, and more, ensuring that the final product meets their specific needs.

6.4.3 Different applications of 3D printing
3D printing has revolutionized many industries by offering cost-effective, customizable, and rapid manufacturing solutions. Here are some of the most significant applications of 3D printing across various fields such as functional prototypes, healthcare products, etc.

1. Functional prototypes
- **Rapid prototyping**: 3D printing allows businesses to create functional prototypes quickly and affordably. Engineers and designers can test design concepts, functionality, and fit before moving to mass production, drastically reducing the development cycle.
- **Iterative design**: 3D printing enables easy design modifications and multiple iterations without significant additional cost, allowing engineers to refine prototypes efficiently.

Example: Automobile manufacturers, such as Ford and Tesla, use 3D printing to prototype new car parts and verify designs before committing to traditional manufacturing methods.

---

2. Healthcare products
- **Prosthetics and orthotics**: Customized prosthetic limbs and orthotic devices can be printed based on the unique anatomy of the patient, offering a perfect fit and

improved functionality at a much lower cost compared to traditional manufacturing methods.

- **Dental and orthodontic devices**: Dentists and orthodontists use 3D printing to create crowns, bridges, implants, and clear aligners that are custom-fit to a patient's mouth.
- **Surgical planning models**: Surgeons can use 3D-printed models of patients' organs or bones to plan complex surgeries more accurately, leading to better outcomes.
- **Bioprinting**: Although still in the research phase, bioprinting aims to create functional tissues and organs using living cells, which could potentially revolutionize organ transplants.

Example: Companies like Open Bionics have developed affordable 3D-printed prosthetic arms for amputees.

3. Aerospace and defense

- **Lightweight components**: Aerospace engineers use 3D printing to manufacture lightweight, complex parts that reduce the weight of aircraft and spacecraft, thereby improving fuel efficiency and performance.
- **Tooling and spare parts**: 3D printing is used to create customized tooling and spare parts on-demand, reducing lead times and storage costs for aerospace and defense manufacturers.
- **Customization**: The ability to print components tailored to specific needs is particularly useful in defense applications where highly customized, mission-critical parts are needed.

Example: NASA uses 3D printing to create rocket engine components, and the International Space Station (ISS) has a 3D printer to produce spare parts in space.

4. Consumer goods and custom products

- **Personalized products**: 3D printing allows consumers to customize products such as jewelry, shoes, phone cases, and eyewear. This on-demand manufacturing leads to unique, tailor-made goods.
- **Home décor**: Artists and designers use 3D printing to create intricate and unique furniture, lamps, and decorations that would be difficult to produce using traditional methods.
- **Fashion**: Designers are exploring 3D printing for creating one-of-a-kind garments, shoes, and accessories. 3D-printed fashion enables highly detailed and complex designs, often with sustainable materials.

Example: Companies like Adidas have used 3D printing to develop customizable, high-performance footwear.

5. Architecture and construction

– **Model making**: Architects use 3D printing to create detailed physical models of buildings and landscapes, allowing clients and stakeholders to better visualize projects.
– **Construction**: 3D-printing technology is being used to print full-scale buildings or components such as walls, using materials like concrete. This speeds up the construction process and reduces labor costs.
– **Customization and sustainability**: 3D printing in construction allows for highly customizable designs, including complex geometries, while reducing material waste and making construction more sustainable.

Example: Several companies are experimenting with 3D-printing houses, with successful projects in the Netherlands, China, and the USA.

6. Automotive industry

– **End-use parts**: Beyond prototypes, 3D printing is being used for producing end-use automotive parts, particularly for limited production runs or highly specialized vehicles.
– **Customization**: It enables manufacturers to create custom car interiors, dashboards, or personalized car components for unique customer demands.
– **Vintage car parts**: 3D printing helps to reproduce discontinued or hard-to-find parts for vintage and classic cars, preserving their historical integrity.

Example: Porsche and BMW use 3D printing for both prototypes and producing spare parts for older car models.

7. Education

– **STEM education**: Schools and universities use 3D printers to teach students about design, engineering, and manufacturing processes. 3D printing also helps to visualize complex concepts in subjects such as biology, physics, and engineering.
– **Hands-on learning**: It allows students to bring their ideas to life, helping them to better understand theoretical concepts through hands-on projects.

Example: Many universities offer design and engineering students access to 3D printing labs to prototype inventions or academic projects.

8. Food Industry

– **3D-printed food**: Companies are experimenting with 3D printing food using edible materials such as chocolate, dough, or plant-based proteins. This technology can customize the shape, texture, and nutritional content of food.
– **Space and remote areas**: 3D-printed food is also being explored for space missions, military use, and providing nutritious meals in areas with limited resources.

Example: Companies like natural machines have developed 3D food printers like "Foodini" to create custom-shaped meals.

9. Art and design

– **Artistic creations**: Artists and sculptors use 3D printing to explore new forms and mediums that would be difficult to achieve through traditional methods. Complex sculptures, installations, and custom pieces are created with high precision.
– **Restoration**: 3D printing has been used to recreate historical artifacts, artwork, and sculptures, aiding in conservation efforts.

Example: Some museums use 3D printing to reproduce rare artifacts for public display, while keeping the originals protected.

10. Electronics and robotics

– **Custom enclosures and cases**: 3D printing allows for the rapid design and production of custom enclosures for electronic devices, robots, or other gadgets. This enables tailored designs for hobbyists and engineers.
– **Wearable technology**: Wearable tech such as smartwatches, fitness trackers, and even medical devices can have customized, ergonomic designs, thanks to 3D printing.

Example: 3D printing is used to produce housings for drones, robots, and IoT devices.

– **Conclusion:** 3D printing's versatility and ability to customize and rapidly produce parts make it an invaluable tool across multiple industries. From healthcare and aerospace to consumer goods and education, 3D printing continues to expand its impact, driving innovation and enabling new possibilities.

Questions

Long answer-type questions

1. What is 3D printing and how is a 3D CAD model converted into a printable file using slicing software? Discuss the importance of slicing parameters such as layer thickness, support structures, and infill pattern in ensuring the structural integrity and functionality of the printed part. **(IES 2015)**

2. What is the role of slicing software in 3D printing? Describe the slicing algorithm and how it translates a 3D model into a series of print layers. Discuss the differences between slicing techniques like adaptive slicing and fixed slicing, and how each affects the print quality and efficiency. **(IES 2016)**

3. Discuss the 3D-printing process and the importance of slicing software in converting a 3D CAD model into a machine-readable format for 3D printers. Explain how slicing software affects print quality, speed, and material usage. Describe the main parameters controlled by the software, such as infill percentage, layer thickness, and print speed. **(GATE 2017)**

4. Explain how Digital Twin technology is transforming product development. Discuss the ways in which it aids in improving design accuracy, reducing time-to-market, and increasing the efficiency of product testing and validation. **(IES 2019)**

5. Discuss the concept and significance of additive manufacturing (AM) in modern manufacturing industries. Provide an overview of different AM technologies, including FDM, SLA, and SLS. Explain how each technology impacts product design, prototyping, and production efficiency. **(IES 2019)**

6. What is a Digital Twin, and how is it used in product development to enhance innovation and speed up time-to-market? Explain how Digital Twin technology supports virtual prototyping and reduces the cost of product testing. **(GATE 2020)**

7. Discuss the application of Digital Twin technology in the aerospace industry. Provide a case study where Digital Twin has been used to simulate aircraft performance, predict maintenance needs, and improve safety. Explain the impact of this technology on the design and maintenance phases of aircraft. **(GATE 2021)**

8. Explain the concept of a Digital Twin in manufacturing systems. How does it enhance the design, monitoring, and maintenance of physical assets? Discuss the applications and challenges in implementing Digital Twin technology in an industrial environment. **(GATE 2022)**

9. Explain the concept of Digital Twin in the context of Industry 4.0. Discuss how this technology is transforming manufacturing and operations by enabling real-time monitoring and predictive maintenance. What are the potential challenges in implementing Digital Twin in industrial applications? **(IES 2022)**

10. Explain the role of additive manufacturing (AM) in the manufacturing of customized parts. Discuss its applications in the automotive and aerospace industries, highlighting how AM enables the creation of lightweight, intricate components. Also, discuss the advantages of using AM for rapid prototyping and tooling. **(IES 2022)**

Short answer-type questions

1. List the materials commonly used in Additive Manufacturing and discuss their mechanical properties that make them suitable for AM processes. **(IES 2016)**

2. Give an example of how Digital Twin has been implemented in industry and explain the benefits it provided. **(IES 2018)**

3. Compare and contrast the FDM, SLA, and SLS 3D-printing technologies in terms of their working principles and applications. **(IES 2019)**

4. Discuss the use of Digital Twin in product lifecycle management and its impact on product development processes. **(IES 2019)**

5. Describe different types of Additive Manufacturing technologies and explain their applications in various industries. **(IES 2020)**

6. Describe the process algorithm used by slicing software in 3D printing. What are the common slicing techniques, and how do they impact the quality of the final print? **(GATE 2021)**

7. Define the concept of Digital Twin and explain its significance in industrial applications. **(GATE 2021)**

8. How does the Digital Twin concept enhance predictive maintenance and operational efficiency in manufacturing systems? **(IES 2021)**

9. What are the main Additive Manufacturing (AM) technologies? Discuss their advantages and limitations. **(GATE 2021)**

10. Discuss a case study where Digital Twin technology was implemented in the automotive or aerospace industry. What were the outcomes? **(GATE 2022)**

Objective questions

1. What is the main purpose of a Digital Twin in manufacturing? **(GATE 2021)**
   (a) To replace physical models with digital versions
   (b) To provide real-time data and predictive analytics for maintenance
   (c) To store all product design data in one location
   (d) To perform simulation in the early stages of product design

2. Which of the following is a key application of Digital Twin technology in manufacturing? **(IES 2020)**
   (a) Reducing material cost by removing physical testing
   (b) Creating digital versions of workers for training purposes
   (c) Enabling real-time monitoring of production systems and predictive maintenance
   (d) Automating assembly lines using robots

3. How does a Digital Twin support product development? **(GATE 2019)**
   (a) By offering real-time feedback on product performance during the design phase
   (b) By replacing the need for physical prototypes
   (c) By performing all product testing automatically
   (d) By conducting market analysis for the product

4. Which industry has widely adopted Digital Twin technology to monitor and optimize operations? **(GATE 2020)**

   (a) Healthcare   (b) Automotive   (c) Retail   (d) Education

5. Which of the following Additive Manufacturing technologies is primarily used for producing parts layer-by-layer from liquid resin using UV light? **(IES 2021)**
   (a) FDM   (b) SLA   (c) SLS   (d) EBM

6. Additive Manufacturing (AM) is particularly useful for which of the following? **(GATE 2019)**
   (a) High-volume production of large components
   (b) Rapid prototyping and creating complex geometries
   (c) Reducing energy consumption in industrial operations
   (d) Large-scale mass production with minimal cost

7. Which of the following materials is commonly used in FDM (Fused Deposition Modeling) for 3D printing? **(IES 2020)**
   (a) Photopolymer (b) Nylon (c) Stainless steel (d) Titanium

8. Which of the following 3D-printing technologies uses powdered material that is fused by a laser? **(GATE 2021)**

   (a) FDM   (b) SLA   (c) SLS   (d) DLP

9. What is an emerging trend in Additive Manufacturing technology?   **(GATE 2022)**
    (a) The use of materials that do not require any post-processing
    (b) Increased use of 3D-printed food items
    (c) Development of multi-material 3D-printing capabilities
    (d) Use of 3D printing only for educational purposes
10. When designing parts for 3D printing, which factor should be considered to ensure that tolerances
    are within acceptable limits?   **(IES 2019)**
    (a) The type of 3D printer used
    (b) The material cost
    (c) The layer thickness and print orientation
    (d) The printing speed

Answers
1(B)   2(C)   3(A)   4(B)   5(B)   6(B)   7(B)   8(C)   9(C)   10(C)

Chapter 7
Manufacturing-related topics

7.1 Material, manufacturing, and socio-economic development

7.1.1 Importance of manufacturing in socio-economic development

The role of manufacturing is vital for the economic and social development of any nation. Approximately 25% to 30% of the value of all domestic products is produced by the manufacturing industry. The manufacturing activity of a nation directly impacts its economy. As manufacturing activity increases in a country, the living standards of its people also improve. Manufacturing is the indicator growth and prosperity of a nation.

Manufacturing involves people from different areas, discipline, and skill sets, which help them work as a team and improve brotherhood. Since manufacturing processes produce a lot of goods by transforming raw materials to create useful products.

The economic aspect is one of the most important considerations in any manufacturing operation. However, this is just one aspect of production. In global market, the price of a product must be competitive with other similar products. The total cost of a product depends on various factors. This total cost can be reduced by improving various manufacturing activities involved in producing that product. The total cost of a product consists of:
1. Material cost
2. Tooling and fixed cost
3. Direct and indirect labor cost

These developments lead to decision-making power with workers, team spirit and mutual trust, and integrity in management and workers.

7.1.2 Importance of materials in socio-economic development

Material is an important input to manufacturing. It plays a vital role in our culture, since pre- civilization era. Civilizations are known by materials, e.g., stone, bronze, steel ages, etc. Manufacturing high-quality products at lower possible price requires an understanding of many factors like product design, selection of material and manufacturing processes etc. Designs are modified to improve product performance. The characteristics of new materials are considered to make manufacturing and assembly easier and at reduced cost. Availability of a wide variety of materials at lower cost is a major challenge. To face these challenges, the knowledge of innovative and creative approaches to design and manufacturing technology is a must. This will enable the products to be available at affordable cost and will improve the standard of

https://doi.org/10.1515/9783112205891-007

living to the common people and the quality of life in the following main areas. For development of any nation by technically or economically, manufacturing play very important role in some field which are directly influenced the nation.

1 **Environment:** Environment has a direct influence on human life. Due to highly polluting environment the humanity is suffering. Material science is discovering new materials which are less toxic and easily recyclable and environment friendly. New energy generation technology is reducing industrial pollution by generating electrical energy using new renewable resources.

2 **Health:** Material and manufacturing are not only making the life more comfortable by providing facilities like refrigerator, air conditioner, travelling transportation, but also contributing to extend it by making artificial heart, valves, safe drug, artificial bones, non-corrosive organs, and making special gadgets for the use of surgeons.

3 **Consumer goods and communication material:** The present era can rightly be called the era of quality, i.e., superior quality, better products at low cost by using newly developed materials. These new materials can be easily fabricated machined or cost to the required shape.

Developments are underway to produce new electronic magnets and optical materials to fulfill the need of faster communication.

7.2 Plant location

A plant is a place where people, materials, and equipment are systematically and logically arranged to manufacture a desired product. The location of the plant can have a significant impact on the overall profitability and success of a project, as well as its future expansion. Several factors need to be considered when selecting a suitable plant site. There are some general considerations like the selection of region, (considering the natural resources and Government policies), followed by selecting the locality within that region (urban, sub-urban, or rural). The next step is the selection of the actual site. Factors influencing the actual site location are as follows:

a) Availability of raw materials
b) Availability of labor
c) Proximity of the market
d) Transportation and communication facility
e) Availability of land (cost, probability of floods, earth quakes, drainage, etc.)
f) Attitude of local community
g) Climate conditions (textile mill require moist climate)
h) Political and strategic consideration
i) Availability of energy/fuel
j) Water

k) Ancillary units around
l) Finance and business facilities
m) Other facilities (hospitals, market, schools, and banks)
n) Taxes bye-laws
o) Government incentives
p) Economic aspects

7.3 Plant layout

A plant layout is an optimum arrangement of physical facilities and services in an existing or proposed plant in order to obtain maximum, output, and highest quality at lowest cost. An optimum layout plan minimizes distances that materials, equipment, and operators need to move, while accounting for the handling equipment required. It also minimizes the energy required for product-material movement against gravity (resistance).

7.3.1 Objectives of a good layout

1. Effective utilization of available floor space (horizontally and vertically).
2. Integrate the production centers into an efficient production unit.
3. Reduce material handling distances and time, and improve decency, and order-lines.
4. Provide flexibility in the layout to permit changes in production programs.
5. Reduce hazards and accidents; take care of the safety and health of labors.
6. Permit better and efficient utilization of labor, machines, and equipment.
7. Job satisfaction for workers and convenience (control temperature, humidity, noise, light, and ventilation).
8. Permit easy supervision and control.
9. Permit easy maintenance.
10. Improve productivity and reduce physical effort.
11. Reduce in process inventory.
12. Remove bottle necks.

7.3.2 Factors affecting plant layout

1. **Existing factory building:** The existing building dimensions provide information about the floor space available for the layout. The existing building design is very important while designing the special requirement as central air conditioning system.

2. **Type and nature of product:** This is a very important factor while designing a plant layout.
    Layout is dependent on the type of product that is to be produced. For a uniform product, a product layout is suitable, whereas for a custom made product, a process layout is more appropriate.
3. **Production process being used:** For assembly line and transfer line in product layout is more suitable. In an intermittent production system, a process layout is best.
4. **Type of machinery used:** Special purpose machine are arranged according to the sequence of operations needed for the product layout, and general-purpose machines are arranged as per the process layout.
5. **Ease of repairs and maintenance:** In light of this consideration, every machine should be positioned such that there is enough room surrounding it for the team and equipment needed for machine repair.
6. **Provision of human needs:** This is a very important factor which affect s plant layout. Toilets, drinking water, lockers, restrooms, and other staff amenities must be properly arranged. If there are any effluents, a proper disposal facility must be provided.
7. **Healthy plant environment:** A factory layout should ensure proper ventilation and heating, a hazard-free working area, adequate safety arrangements, support for workers' health and safety, smooth movement of people and materials, and consideration for future growth and diversification.

7.3.3 Types of plant layout

Plant layouts can broadly be classified of four types:
i) Product or line layout: This layout is also called a line layout. Here the operations are performed in the sequence of their occurrence and the equipment's is also arranged in the same sequence. This layout is used for the continuous production of a given product, as shown in Figure 7.1

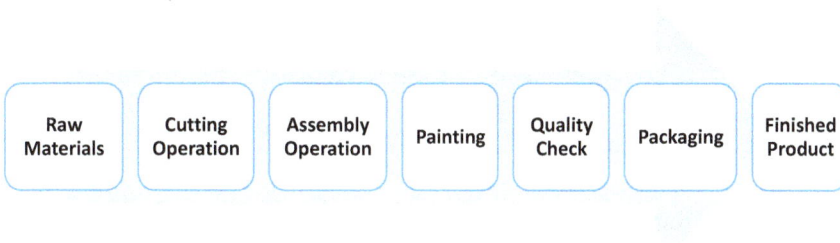

| Raw Materials | Cutting Operation | Assembly Operation | Painting | Quality Check | Packaging | Finished Product |

Figure 7.1: Product or line layout (arranged in sequence of operations).

- **Advantages**
  a) Reduced product movement and processing time.
  b) Simple production planning, better coordination, and control
  c) Reduced space requirements for a given production volume.
  d) Automatic material handling possible
  e) Reduced in-process inventory (less material in-process)
  f) Smooth work-flow
  g) Easy for workers to acquire (or improve) necessary skill.
- **Limitations**
  a) Layout is not flexible, and product changes require major changes to the layout.
  b) The pace of work depends on the pace of the slowest machine.
  c) More machines are needed as the machines are arranged operation wise. Standby machines are needed as one machine failure may result in complete shutdown of production line.
  d) The system requires a higher capital investment.
  e) Inspection becomes difficult.

ii) Process layout: A functional layout, also known as a process layout, groups similar function machines in one location (e.g., all lathes together) and is used for job-order or non-repetitive production, as shown in Figure 7.2, which represents a workshop with different sections for specific processes.

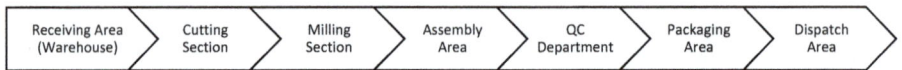

Receiving Area (Warehouse)	Cutting Section	Milling Section	Assembly Area	QC Department	Packaging Area	Dispatch Area

Figure 7.2: Line diagram representing a process layout with different sections for specific processes.

- **Advantages**
  Process layout provides the following advantages:
  a) High flexibility in assigning work to equipment and workers.
  b) Equipment utilization is high.
  c) Comparatively smaller number of equipment needed.
  d) No monotony of job exists as the jobs keep changing.
  e) Workers are safe from radiations, heat, and hazards.
- **Limitations**
  a) Production control is difficult.
  b) High work-in-process inventory.
  c) Back tracking causes raw materials to travel longer distances during processing.
  d) Consumes more space.
  e) Needs more inspections and better co-ordination.

f) Automatic material handling is extremely difficult.

g) Processing time increases.

iii) Combination layout: It is also known as a hybrid layout. This layout has the advantages of both the product and process layouts. In industries, pure product or process layout are rare. A combination layout becomes more relevant in situations where a product is made in different shapes and sizes. The machines are arranged in process grouping and these groups are arranged in process groupings and these groups are arranged in a sequence to manufacture various types and sizes of products (see Figure 7.3).

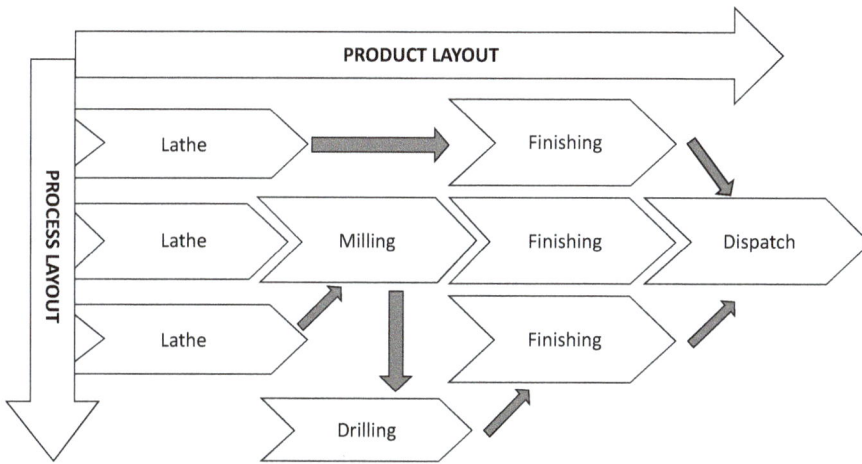

Figure 7.3: Line diagram of a combination layout.

iv) Fixed layout: For the manufacture of large structures like ships, submarines, bridge, spacecraft, this type of layout is the only alternative, as the product cannot be moved, whereas the men, materials, and machines are moved around the product, as shown in Figure 7.4.

– **Advantages**

a) A single or a group of workers could be assigned to a project from beginning to end.

b) Maximum flexibility.

c) Least material movement.

d) Many different projects could be taken up with the same layout.

– **Limitations**

a) Needs a highly-skilled work force.

b) Tools and machines require more time to reach the location.

c) Low labor and machine utilization.

d) Capital investment is very high, and the production period is long.
e) Needs a very large space for material storage and equipment around the product.

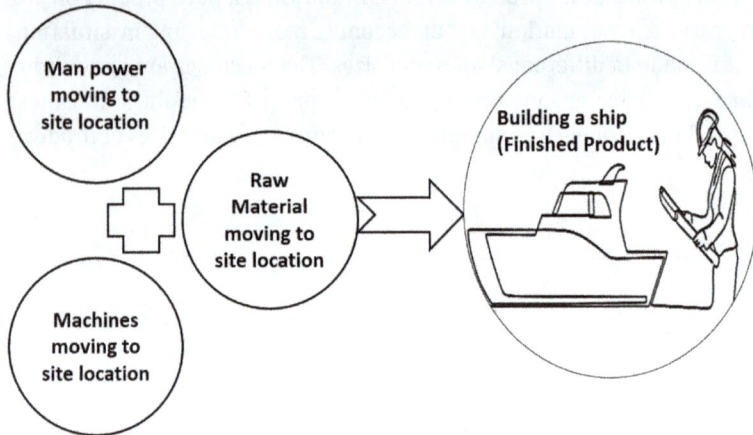

Figure 7.4: Line diagram of a fixed or static product layout.

7.4 Production

It is a process of converting raw materials into finished products with the application of tools and techniques. Or we can say, the systematic operations done on raw material to produce desired product or output. Production refers to absolute output, i.e., the number of parts produced.

Table 7.1 with figure 7.5, represents the relationship among volume and variety of products in various production system.

– **Types of production**
   i.   Job-type production
   ii. Batch production
   iii. Mass production

i. Job-type production: In this type of production, the products produced are of high variety and low volume. The production unit generally produces one or a few special designed product. In this category, products are produced as per customer specification.

– **Characteristic of job-type production**
   a) Produced in low volumes with a high variety of products.
   b) Utilizes general-purpose machines and tools to support diverse operations.
   c) Material flow is irregular due to varying work content at different stations.

d) Requires a highly skilled workforce with close supervision.
e) Involves a large work-in-progress inventory.
f) Needs a flexible material-handling system to transport objects of different sizes along diverse paths.
g) Managing numerous components with high variation poses challenges in planning, scheduling, and coordination.

ii. Batch production

Batch production involves manufacturing products with similar designs but varying sizes and capacities in batches. Each batch is produced at regular intervals and stored in warehouses until sold. Examples include turbines, pumps, motors, and pharmaceutical products of different types and capacities.

– **Characteristics of batch production**
   a) Production runs (batch sizes) are short.
   b) Machinery and plant setups are adjusted for a specific batch and reconfigured for the next.
   c) Frequent setup changes are required.
   d) Workers must have specialized skills for specific manufacturing operations.
   e) Requires less supervision compared to job-type production.
   f) Flexible machinery, plant setup, and material handling systems are essential.
   g) High work-in-process inventory.
   h) Shorter cycle time compared to job-type production.

iii. Mass production

Mass production is defined by high-volume, low-variety manufacturing, where standardized products are produced in large quantities and stored in warehouses for dispatch. Common examples include automobile and bearing manufacturing.

– **Characteristics of mass production**
   a) Continuous material flow.
   b) Requires special-purpose machines with high production rates.
   c) Mechanized material handling (e.g., conveyors).
   d) Low-skilled labor is sufficient.
   e) Shorter cycle time.
   f) Low work-in-process inventory is due to line balancing.
   g) Limited machine flexibility.
   h) Higher raw material inventory is required.

Table 7.1: Example of volume-variety relationship for different types of production system.

Production systems	Variety	Volume	Example
Project production	High	Low	Construction of buildings, shipbuilding
Jobbing production	High	Low to medium	Custom furniture, specialized tools
Batch production	Medium	Medium	Bakery items, clothing production
Mass production	Low	High	Automobiles, electronics
Continuous production	Very low	Very high	Oil refining, chemical production

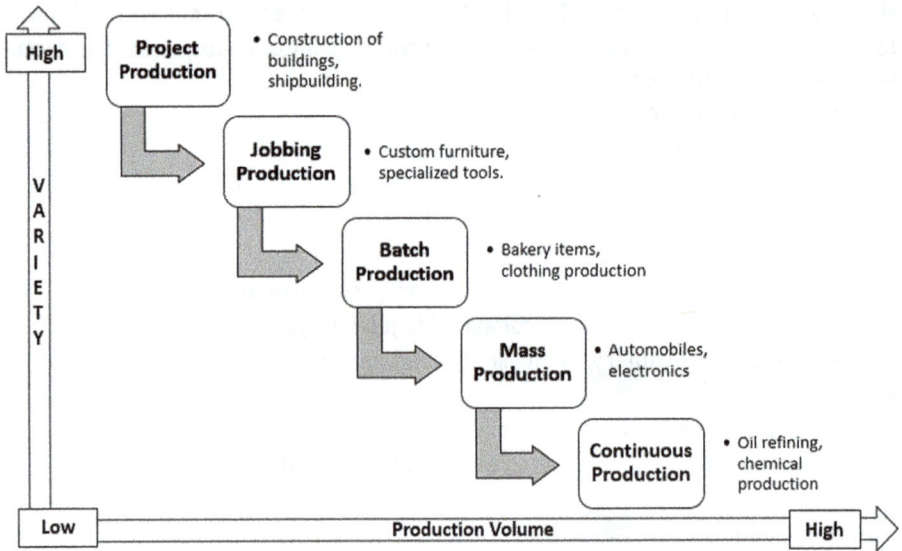

Figure 7.5: Volume-variety relationship with example for different types of production system.

7.5 Production vs. productivity

– Production is the process of transforming raw materials into finished products by executing a series of manufacturing operations in a predetermined order.
– Production refers to the total output, meaning that if the input increases, the output typically increases proportionally, keeping productivity constant. However, productivity improves when output increases without additional input or when the same output is achieved with fewer resources.
– Production is measured in terms of monetary value or the number of units produced, without considering resource input. In contrast, productivity reflects an approach focused on maximizing output with minimal resource input, ensuring that the benefits of increased efficiency reach a broader audience. The goal is to achieve optimal resource utilization by adopting new technologies and methods.

– A summary of comparision between production and productivity has been represented in Table 7.2

Table 7.2: Comparison of production vs. productivity.

Aspect	Production	Productivity
Definition	The total output of goods or services produced in a given period	The efficiency of production, measured as output per unit of input (time, labor, or resources)
Focus	Quantity of output	Efficiency and effectiveness of production
Measurement	Measured in absolute numbers (e.g., units produced, tons manufactured)	Measured as a ratio (e.g., units per hour, output per worker)
Improvement factors	Increased resources, machinery, or workforce	Better technology, skills, workflow optimization, or automation
Goal	Meeting demand by producing more	Maximizing efficiency with minimal resources
Example	A factory produces 1,000 cars in a month	The factory improves efficiency, producing the same 1,000 cars using 10% fewer resources

7.6 Non-metallic material

7.6.1 Common types and uses of wood

Wood is an organic material that is found in the natural composite of cellulose fibers in a matrix of lignin. It has high-compressive strength. Cellulose fibers are very strong in tension. When a tree is living, wood performs supports which help to grow large. Wood also plays a very important role to mediate the transfer of water and nutrients from root to leaves. There are various advantages of using wood. Some of them are given below:

– **Advantages:**
   a) Easy to work in order to get desired shape and size.
   b) Lighter in weight and stronger.
   c) Good response to polishing and painting.
   d) Non-conductor of heat.
   e) It has good compressive strength, durability, lightness, and economical as compared to other material.
   f) Cheaper than other material.

1. Hard and soft woods
Hardwood is obtained from trees with broad leaves, while softwood is obtained from those trees having needle shaped leaves. The main difference between softwood and hardwood are given below in Table 7.3.

Table 7.3: Comparison of hardwood with softwood.

Hardwood	Softwood
It has good refractoriness	It may catch fire soon because of low refractoriness
Easy to be worked	Difficult to be worked
Good tensile and shear resistance	Good tensile strength but low shear resistance
Light in color	Dark in color
Contain a good amount of acid, it is a non-resinous wood	Fragrant smell and regular texture
Fibers are close and compact	Straight fibers and fine texture
Heavy in weight	Light in weight
Growth rate is slower	Growth rate is faster
Due to high density, they are hard in nature	Due to lower density, they are soft in nature
Example: teak, oak, birch, maple, etc.	Example: Spruce, larch Douglas-fir, etc.

7.6.2 Seasoning of wood

Seasoning is the process of removing or reducing excess or unwanted moisture from wood. If the unwanted moisture is not removed, it can cause uneven shrinkage, warping and twisting. The main objective of seasoning is to remove the moisture in to make it suitable for various purposes. There are various advantages of seasoning; some of them are given below.

a) After the removal of moisture, hardness and durability increased
b) Shock resistance increase
c) No chances of shrinkage, twisting, or warping after seasoning
d) More response to polishing and painting
e) Resistance to fire is increased

1. **Air seasoning:** It is a method of removing moisture from wood by exposing the wood to open air. The wood is not directly exposed to sun or rain. In this process, logs of wood are stacked in a shed to allow free circulation of air through them. For air seasoning a platform of around 30 to 40 cm high is constructed on the ground. A layer of sand is placed over the top surface of the platform to avoid the moisture from effecting the wood from the bottom. The shed is upright over the platform and logs of wood was stacked as shown in Figure 7.6. The wood is kept in this shed for long period of time under various temperature conditions. The stacked logs of wood should be periodically turned upside down to accelerate the

removal of moisture. Excess moisture in the wood evaporates due to the free circulation of air through the stacked wood.

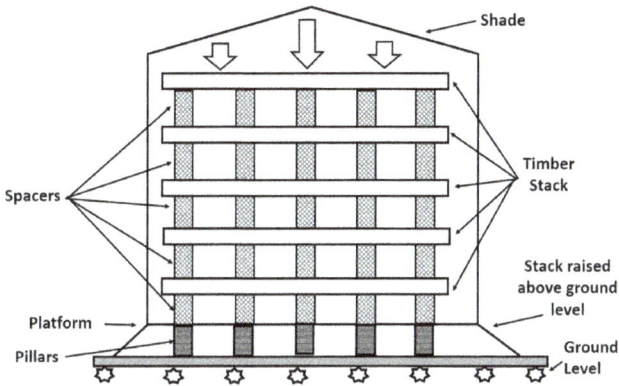

Figure 7.6: Stacked wood for air seasoning.

2. **Artificial seasoning:** It is also called Kiln seasoning, as shown in Figure 7.7. It is a method of removing moisture by keeping the log of woods over a large trolley, which is driven into hot chamber (Kiln). In this kiln, the wood is kept at control temperature and humidity for nearly a night. Exact time is for drying depends upon the moisture present in wood and the quality of wood. When hot air is allowed to flow in chamber, the temperature of the chamber rises gradually, leading to drying and evaporation of the wood, starting with a low rate and gradually increases with increase in temperature.

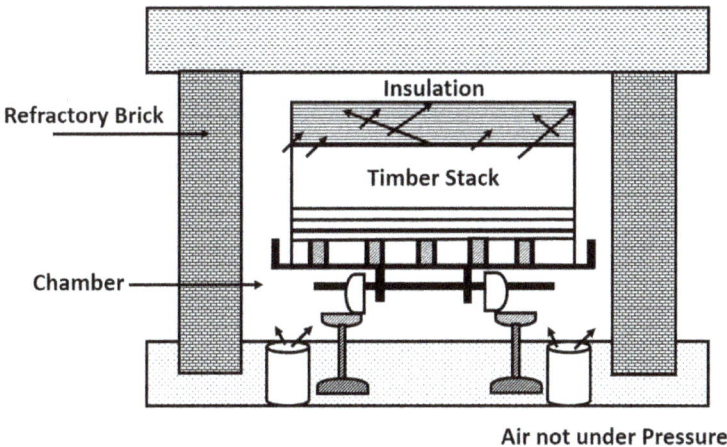

Figure 7.7: Natural seasoning (kiln seasoning).

Apart from these methods of seasoning there are some more methods are available to remove moisture content from wood:
a) Water seasoning
b) Chemical seasoning
c) Electric seasoning
d) Combination of air and Kiln seasoning

7.6.3 Cement concrete

Concrete is a very hard material that is used to withstand high compressive loads. Generally, concrete is used in the construction of roads, building, and bridges that need to be strong and durable. Concrete is manufactured by combining sand or gravel and a binder with chemical additives and water. When the mixture dries, it becomes harder and develops properties like stone. All the steps in making cement concrete are shown in Figure 7.8.

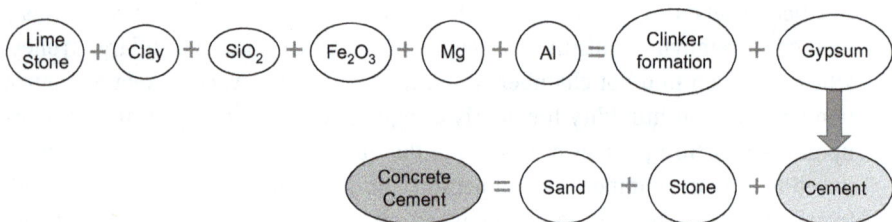

Figure 7.8: Steps in making cement concrete.

Making cement concrete involves several steps, from selecting and preparing raw materials to curing the final product. The detailed process is explained below:
i.   **Selection and preparation of raw materials:** Cement is made from different material like limestone, silica, and clay.
   a) Cement: Usually Portland cement.
   b) Aggregates: Fine aggregates (sand) and coarse aggregates (gravel or crushed stone).
   c) Water: Clean and free from impurities.
   d) Admixtures (optional): Chemicals added to modify the properties of concrete.
ii. **Proportioning:** Determine the proper mix ratio of cement, aggregates, and water to achieve the desired strength and durability. Common ratios are 1:2:4 (cement: sand) or 1:3:6.
iii. **Preparation:** First of all, make a fine powder of raw material in definite proportion and then heat the material to a temperature at which moisture and other

impurities are removed. After this, the material is burned in a rotating chamber at temperature 1,417 °C. This helps to fuse the material together to form clinkers.

iv. **Mixing:** Clinker is then ground to fine powder, mixed with gypsum, to form cement. Cement is then combined with aggregate (stone and sand) and water.
   a) Batching: Measure the exact quantity of each material.
   b) Mixing: Combine the materials. This can be done manually for small projects or by using a concrete mixer for larger batches. Ensure a uniform mix.

v. **Transporting:** Transport the mixed concrete to the construction site. This should be done quickly to prevent initial setting.

vi. **Placing:** Pour the concrete into the prepared formwork or molds. Ensure that it is placed as close as possible to its final position to avoid segregation.

vii. **Compacting:** When the mixture is ready, it is then placed for compaction. Compaction is done to prevent separation of raw materials, remove air bubbles, and eliminate voids by compacting the concrete. This can be done by using hand tools or mechanical vibrators.

viii. **Finishing:** Smooth and finish the surface of the concrete. Tools like trowels, floats, and screeds are used to achieve the desired finish.

ix. **Curing:** Curing of compacted cement is necessary in order to avoid rapid drying, which may cause crack formation in the cement. Keep the concrete moist to allow it to achieve its full strength. Curing can be done by:
   a) Covering it with wet burlap or plastic sheeting.
   b) Spraying with water.
   c) Applying curing compounds.

x. **Removing formwork:** Once the concrete has gained sufficient strength (usually after 24–48 h), remove the formwork.

xi. **Quality control:** Test samples to ensure the concrete meets the required specifications. Tests include:
   a) Slump test (workability).
   b) Compressive strength test.

7.6.4 Ceramics

Ceramics are a mixture of various metal and non-metallic elements. There are various materials like clay, kaolin, etc. that are processed with high temperature to form ceramics. Clay was the oldest material which was used to produce ceramics. Ceramics have the capability of withstand high temperature (1,000° to 1,800 °C). Ceramics are basically of two types:
a) Traditional ceramics
b) Advanced or modern ceramics

Traditional ceramics include cement, clay product, silicate glass, etc. and advanced ceramic consist of silicon carbide (SiC), pure oxide as Al_2O_3, nitrides (SiN_4), etc.

– **Advantages of modern/advanced ceramics**
  a) They are extremely hard in nature.
  b) Excellent resistance to plastic deformation.
  c) Thermal conductivity is very poor.
  d) Electrical conductivity is also very poor.
  e) Excellent corrosion resistance.
  f) Good wear resistance.

– **Application:** Ceramics have very wide applications in various fields. Some of them are given below.
  a) **Computers/communications:** Insulators, capacitors, microelectronic packaging, fiber optic/laser communications, TV and radio components etc.
  b) **Aerospace:** High-temperature glass windows, fuel cells, space shuttle tiles, etc.
  c) **Military:** Missiles, sensors, air vehicles, etc.
  d) **Medical:** They are called as bio ceramics, which are used in dental restoration, bone implants, etc.
  e) **Automotive:** Spark plugs, piston rings, thermostats, etc.

7.6.5 Rubber or elastomers

Rubbers are linear polymers, are also known as elastic polymers or elastomers. They give an extremely large amount of elastic deformation under an applied force. After removal of load, rubbers regain large deformation rapidly. Rubbers are mainly of two types:
1. Natural rubber
2. Synthetic rubber

1. **Natural rubber:** Natural rubbers are elastic hydrocarbon polymers produced by some plants in the form of latex (milky colloid). A carving (slit) is made on the bark of the tree; from this carve viscid milk color latex sap is collected, after refining this latex it becomes usable rubber. Purified rubber is nothing but a chemical polyisoprene. Natural rubbers are highly flexible, stretchy, and extremely waterproof.
2. **Synthetic rubber:** Synthetic rubbers are artificial elastomer polymers. Chemical polyisoprene can be made synthetically, hence called synthetic rubber. Elastomers can undergo more elastic deformation under external load than any other materials, and it can regain its shape and size without any permanent deformation.

– **Applications**
1. Provide protection against corrosion and abrasion.
2. Used as insulation for shock and vibration, e.g., tires, foot wear, gasket seal, etc.
3. Good insulator for electricity.
4. Used on high friction surfaces.

7.6.6 Composite material

As the word "composite" implies combination of different things. Here, composite materials are the combination of two or more than two different materials (with different properties). When materials with different properties are mixed together at micro or macro level, a third type of material is formed, known as composite material. Composite materials have the properties which are greater than parent materials. Constituents (parent material) of composites are bonded together by mechanically or metallurgical. Composites are mainly of four types, which are discussed below:

1. Agglomerated composites: When a composite is formed by mixing of two agglomerates with some external binders is called an agglomerated composite. Agglomerates are those materials that cannot bond together unless a third binding material is mixed with it. For example, sand and stone are the agglomerates, which are combined by introducing cement in it, and formed concrete cement (CC). Therefore, concrete cement is an agglomerated composite.

2. Laminated or layered composites: When layers of material are bonded together, is formed laminated composite (see Figure 7.9). They also include thin coatings, cladding, laminates, sandwiches, etc. The most famous example of this category composites are plywood. In this type of composites thin layers of wood are bonded with their grain orientation. Bonding of thin layer of wood improves the strength and fracture resistance of plywood. Chances of shrinkage and swelling are reduced.

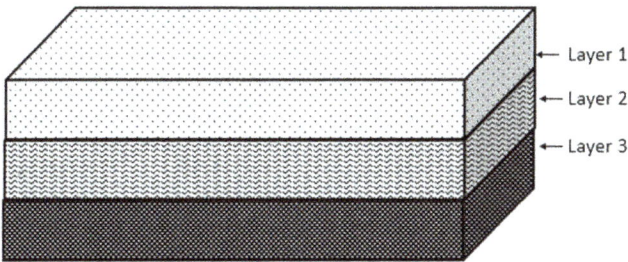

Layer 1
Layer 2
Layer 3

Figure 7.9: Laminated composite.

The diagram illustrates laminated composites, showing multiple layers bonded together for enhanced strength and durability, used in aerospace, automotive, and construction applications.

3. Fiber reinforcement composites: In this category of composites fibers of one material are reinforced in the matrix of another. Reinforcing material is usually soft and ductile in order to enhance stiffness and fatigue strength. Types have been shown in Figure 7.10. For example, reinforced concrete cement (RCC) is a double composite. It consists of agglomerated composite (concrete cement) with reinforced steel bar (fibers). In RCC, agglomerated composite (concrete cement) provides higher compressive strength whereas reinforced steel bar increase stiffness and bending strength of combined structure.

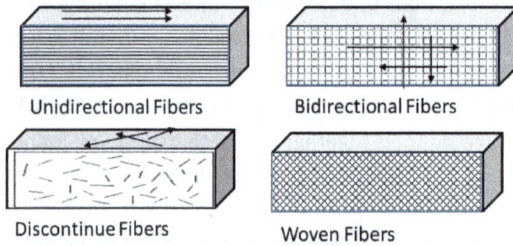

Unidirectional Fibers   Bidirectional Fibers

Discontinue Fibers   Woven Fibers

Figure 7.10: Types of fiber reinforced composite.

The diagram illustrates types of fiber-reinforced composites, including polymer, metal, and ceramic matrices with fibers, used in aerospace, automotive, and construction for high strength and lightweight applications.

4. Coated composite: In this type of composite, the surface of material is coated in order to improve corrosion resistance, wear resistance and to improve the aesthetic of material. Various types of coatings are used, such as metallic coating, inorganic coating, and organic coating. The example shown in Figure 7.11 represents a view of a coated composite material of a specimen coated by paint, which fills the irregularities on the surface of specimen that reduces stress concentration on surface to increase fatigue strength of the specimen material.

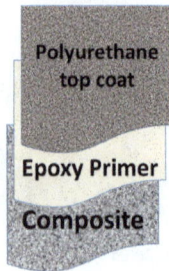

Polyurethane
top coat

Epoxy Primer

Composite

Figure 7.11: Example of coated composite.

7.7 Powder metallurgy process and application

7.7.1 Introduction to powder metallurgy

Powder metallurgy is the process of producing fine metal powder and using this to make a product or component. The component may be produced from individual fine powder or mixed with non-metallic constituents. The component is formed by pressing the powder into a desired shape called mold. After pressing the powder, the component is heated at the temperature below the melting point of the main constituent of the powder. This whole processing is called powder metallurgy and is shown in Figure 7.12.
Powder metallurgy consists of the following steps:

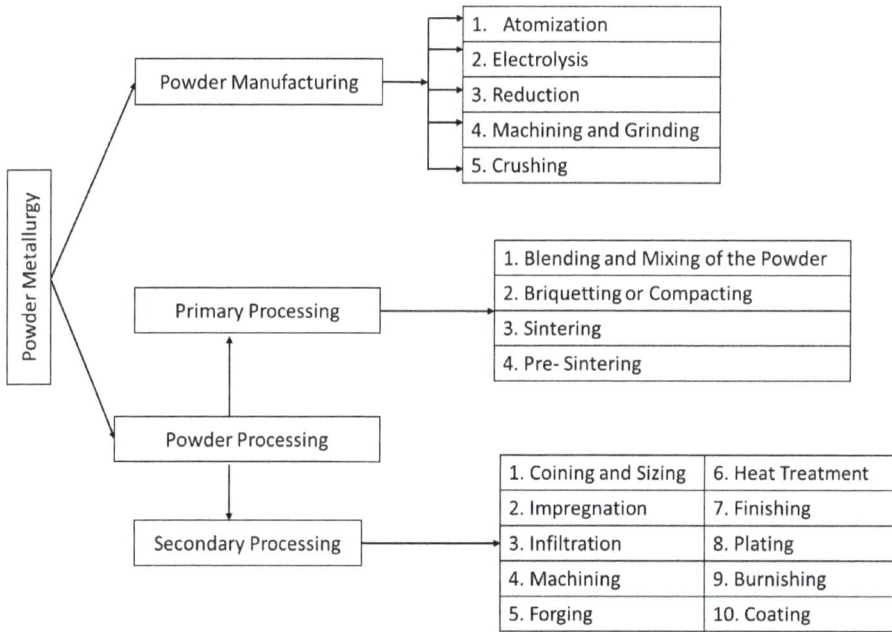

Figure 7.12: Steps involved in powder metallurgy.

7.7.2 Powder manufacturing

In the process of powder metallurgy, the role of metal powder is very important. First, the metal powder needs to be manufactured. Various types and grades of powder are available, which help produce variety of products that meet desirable performance characteristics. The following methods are used to produce metal powder:

1. **Atomization:** In this method, the metal is first melted and then passed through an atomizing nozzle. A very fine stream of molten metal is broken into small droplets by striking the water jet to the stream. These fine droplets are collected at the base of the atomizing chamber. The size of the droplet can be control by controlling the process variable like stream velocity, temperature, etc. Schematic Figure 7.13 shows the manufacturing of powder by atomization method. This process is widely used in the industry for powder manufacturing.

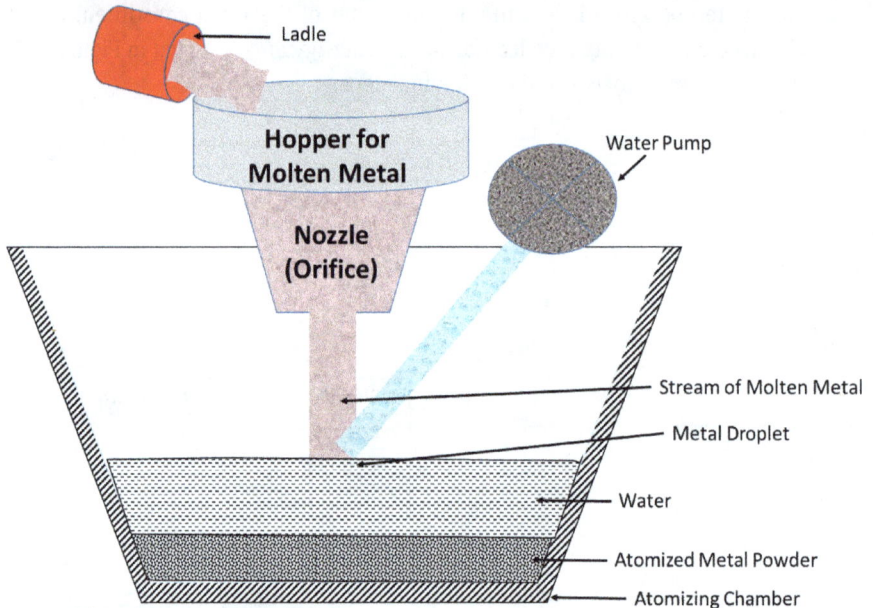

Figure 7.13: Atomization.

2. **Electrolysis:** This technique is employed to produce metal powders such as iron, copper, and silver. It requires a suitable electrolyte for the process. In the production of iron powder, a steel plate serves as the anode, while stainless steel acts as the cathode. These electrodes are connected to a high-powered direct current (DC) source. After approximately 45–50 h, an iron powder deposit with a thickness of about 1.8 mm to 2 mm is formed. The deposit is then washed, screened, stripped, and sized. The resulting iron powder is initially brittle, so an annealing process is carried out to enhance its ductility.

3. **Reduction:** It is a process of producing metal powder using reducing agents. In this method, metal oxide is reduced by reducing gas (like hydrogen gas). When iron oxide is crushed and it is passed in hydrogen gas environment at temperature around 1,045 °C pure iron powders is obtained (in form of sponge like structure).

Apart from iron some other metal powder can be produced like tungsten, cobalt, molybdenum, and nickel.

4. **Machining and grinding:** There are various machines such as grinders, rotator mills; crushers, etc. are used to break down the brittle metal into fine pieces of powder. There is a drawback of this method is that irregular-shaped particles are produced.

7.7.3 Powder processing

It consists of two types of processes:
1. Primary processes
2. Secondary processes

1. **Primary processes**
i. **Blending and mixing of powders:** First, metal powder are mixed and blended. The powder must be mixed in a dry or wet ball mill. In wet mixing, water or any solvent is mixed with the powder, to provide better mixing. Wet mixing is generally practiced as it reduces dust, explosion, and surface oxidation. Blending and mixing involve combining metal and non-metal powders to impart special properties. Proper mixing ensures a uniform particle size distribution.
ii. **Briquetting/compacting:** After the first operation (blending and mixing), the now mixed powder is allowed to pressed. For this operation, a cylinder is used which is open at both the end. From both the end piston applied pressure on mixed powder, and they get compacted. Compression from both end provide better distribution of pressure on object. This process produces the shape of the object near about to final object. This process imparts the desired level and type of porosity and adequate strength for handling. The compacting pressure range is 20 MPa to 1,400 MPa. Compacting pressure can be applied by die pressing, roll pressing and extrusion pressing.

2. **Secondary and finishing operation**
Some operations are needed on sintered parts in order to improve their properties or to achieve special characteristic. There are some operations which are carried on sintered parts, which are as follows:
i. **Coining and sizing:** This operation is done to achieve higher dimensional accuracy and greater strength to the sintered part. This is achieved by high pressure compacting
ii. **Forging:** Forging is a manufacturing process applied to achieve exact shape, fine and uniform grain size, good tolerance and surface finish of the sintered part. Here the metal is shaped by applying compressive forces. The metal is usually heated until it becomes malleable and then deformed using hammers, presses, or dies to achieve the desired shape. This process enhances the metal's structural

integrity, resulting in stronger and more durable components. The types may consist of hot or cold forging, open-die forging, closed-die forging (impression-die forging), roll forging, as per requirements.

The tooling and operational arrangements involved in powder metallurgy and metal forming are depicted in the following figures: Figure 7.14 shows the piston and cylinder setup used for compacting metal powders; Figure 7.15 illustrates a

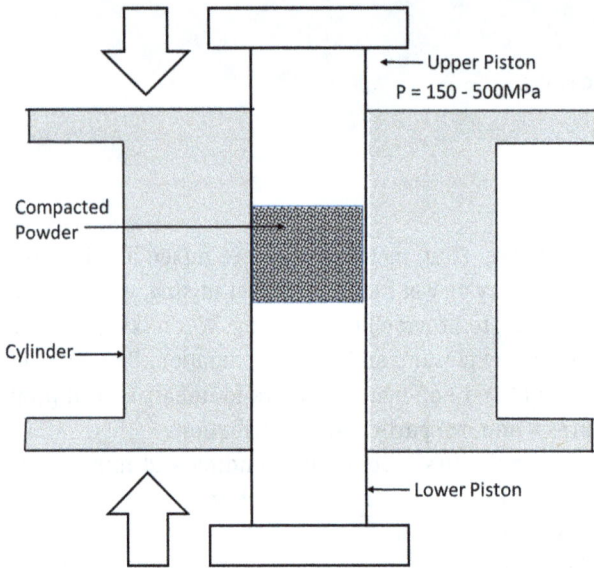

Figure 7.14: Piston and cylinder arrangement for compacting metal powder.

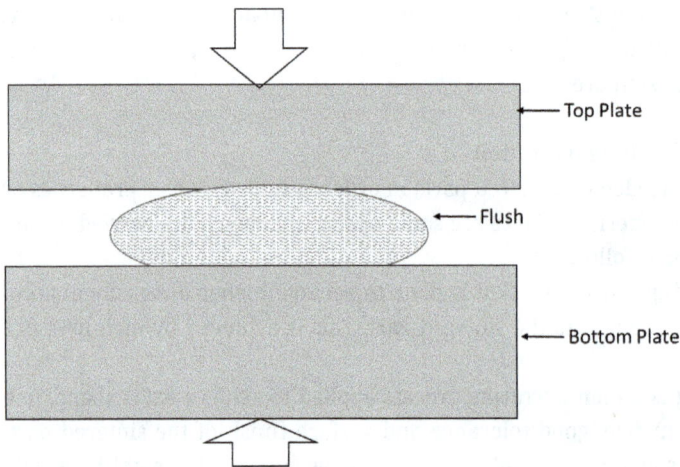

Figure 7.15: Basic forging operation.

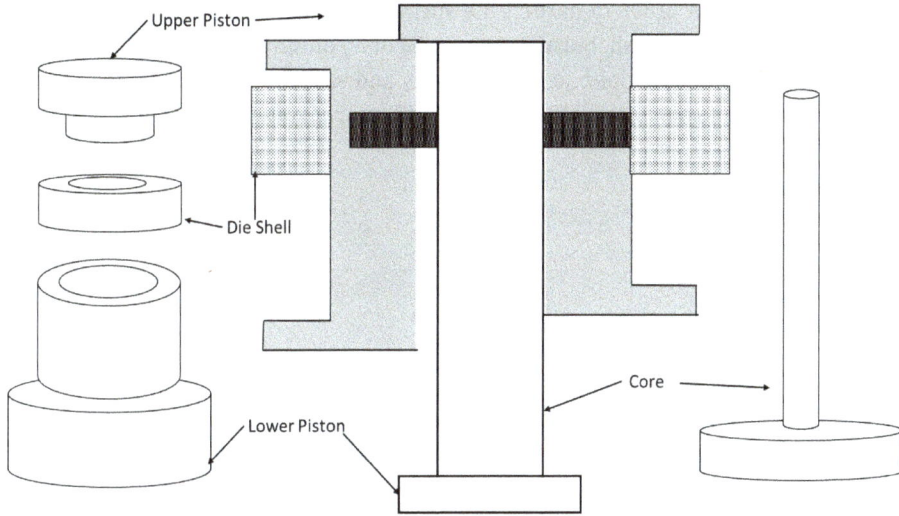

Figure 7.16: Tooling arrangement for briquetting.

basic forging operation; and Figure 7.16 presents the tooling arrangement for the briquetting process.

iii. **Pre-sintering:** pre-sintering is defined as the process in which green compact is heated to a temperature below sintering temperature to increase its strength. This process also removes the lubricants and binder which is added during blending operation.

iv. **Sintering:** Sintering is the process of heating a compacted object below melting point of major constituent in an inert atmosphere and pressing it. Mostly sintering is done at 70% to 30% of the melting temperature. Sintering time varies from half hour to several hours. Sintering is done to improve strength and hardness of final product.

v. **Impregnation:** A completes saturation of sintered part is done by impregnating them with oil or grease.

vi. **Infiltration:** In order to improve hardness and tensile strength, pores of sintered part are filled with some low melting point metal.

vii. **Heat treatment:** Sintered part may be heat treated in order to improve its mechanical properties.

viii. **Machining:** In order to get various geometry of sintered part, various machining operation can be used as milling, drilling, turning, threading etc.

ix. **Finishing:** To improve the surface, we can use all finishing operation on it like coating, burnishing, plating etc. all finishing operations are applicable on it.

x. **Plating:** In order to improve the resistance to wear and corrosion, sintered part may be plated by electroplating.

xi. **Burnishing:** In order to harden the surface or improve the dimensional accuracy and surface finish, burnishing may be done on sintered part.
xii. **Coating:** Coating is done to fill the pores and seal the entire reactive surface. This is done to avoid environmental degradation of sintered parts.

7.8 Plastic product manufacturing

7.8.1 Introduction

Plastics are polymeric materials composed of large molecular structures. In recent times, they have been increasingly utilized in diverse applications, often replacing metals and ceramics. From everyday items like toys and utensils to complex precision components such as heart valves, plastics have not only enhanced convenience but also contributed to improving human life.

Plastics offer a unique combination of beneficial properties, including a high strength-to-weight ratio, excellent resistance to corrosion in various environments, affordability, and the ability to achieve a smooth finish.

1. **Features of plastics**
    i. Plastic is strong
    ii. Light, highly dielectric
    iii. Resistant to chemicals and durable.
    iv. It gives good dimensional tolerances
    v. Excellent surface finish, absorbs vibrations and sound
2. **Limitations**
    i. Low strength
    ii. Low heat resistance
    iii. Soft and less ductile.
    iv. Some plastic is flammable and deform in sunlight

7.8.2 Types of plastic

All plastics are broadly classified into two main groups: thermoplastics and thermosetting plastics, based on their behavior under heat. Their key differences in terms of structure, properties, and applications are summarized in Table 7.4.

i. **Thermosetting plastics:** They are also called thermosets. Thermosetting plastics are like eggs, once heated, they set and cannot be reshaped. These plastics are hardened by heat leads to non- reversing chemical changes.
ii. **Thermoplastics:** Thermoplastics are like wax that can be shaped by heating and reshaped by reheating. They can be reshaped multiple times and on cooling they become hard

Table 7.4: Comparison of thermosetting and thermoplastics.

Thermosetting plastics	Thermoplastics
Thermosets have light meshed branched molecular structure.	The molecular structure of thermoplastics can be either linear or branched.
Curing occur during shaping.	They are flexible at ordinary temperature.
Once shaped, they cannot be re-shaped by heating again.	At 120 °C to 180 °C, they become pasty.
Can be shaped by machining.	Can't be shaped by machining.
Used for making light switches.	Used mostly for packaging, molded furniture, household items, and toys.

7.8.3 Plastic processing method

There are various methods of producing plastic products like; compression molding, transfer molding, injection molding, extrusion molding, blow molding, casting, slush coating, calendering, laminating, thermoforming, etc. Some of the most important and commonly used molding methods have only been discussed below, like; compression, transfer, injection, extrusion, and blow molding methods.

i. **Compression molding:** In this process, preheated plastic resin, either in granule form or as putty-like masses, is placed into an open, heated mold cavity. The mold is then closed using a plunger, and pressure is applied to the upper die to ensure the material fully occupies the mold cavity. Heat and pressure are maintained throughout the process until the material is fully cured. Once curing is complete, the plunger is lifted to eject the finished product, and the upper die moves upward, as illustrated in Figure 7.17.

ii. **Transfer molding:** In this process, the plastic resin is kept in an open mold, is preheated and loaded into a holding chamber called a pot. The material is then transferred or forced in to the preheated mold cavity by plunger through the sprue. The mold is remains closed until plastic resin is cured. This method avoids the turbulence in mold cavity as it was in compression molding, leads to good surface of product. Schematic diagram are shown below which explain the process clearly. In transfer molding method, as shown in Figure 7.18, having a less flash are formed because excessive material move along the parting line of the mold.

The diagram illustrates the transfer molding process, in which heated material is forced into a mold cavity for precise shaping. This process is commonly used in the manufacturing of electrical components, seals, and composite parts

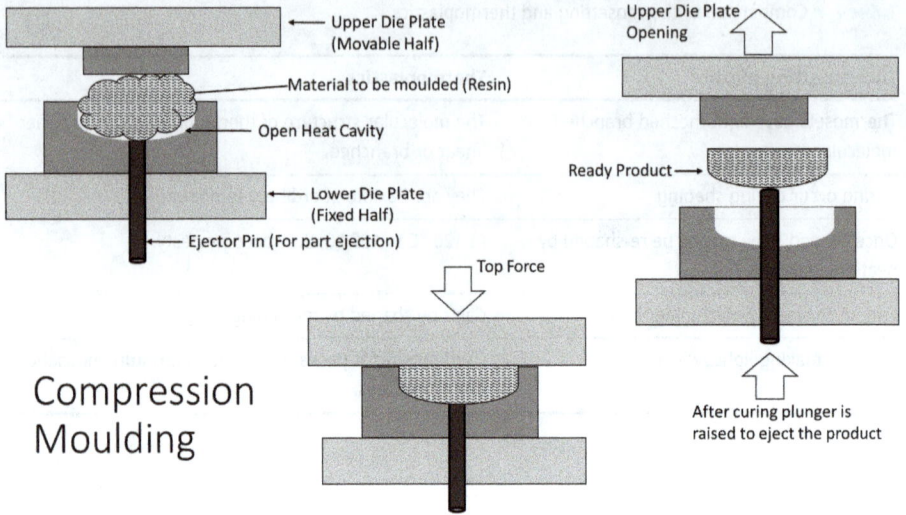

Figure 7.17: Compression molding process.

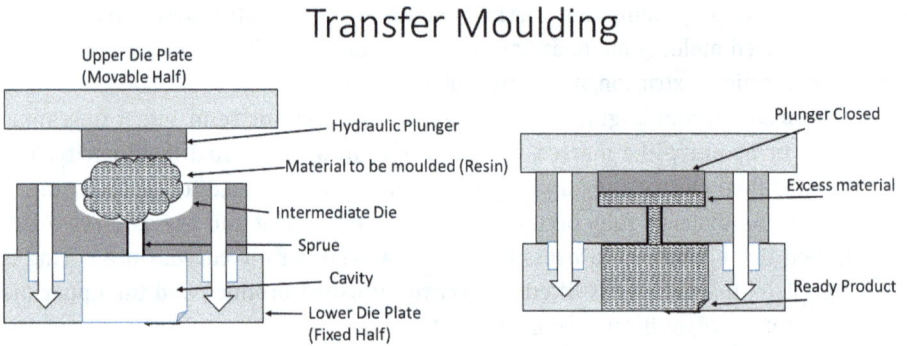

Figure 7.18: Transfer molding.

iii. **Injection molding:** Injection molding machines are similar to die casting machines and are most successfully and commonly used for molding thermoplastics. Granulated material is fed from a hopper in measured amounts to the heating chamber, where it changes into a constituent fluid. The softened plastic is then injected into the mold cavity. The machines are described by the die clamping force in Newton's and by the amount of material in kg/cycle. A schematic diagram of the injection molding machine is shown in Figure 7.19(a) and the process is shown in Figure 7.19(b).

iv. **Extrusion molding:** Thermoplastic materials can be extruded through dies to form simple shapes of any length, as shown in the figure. The process begins with feeding powdered or granulated material into a hopper, where it is heated in a chamber until it transforms into a thick, viscous mass. This molten material is then forced through a die and cooled using air, water, or a chilled conveyor surface, producing long tubes, rods, and specialized sections.

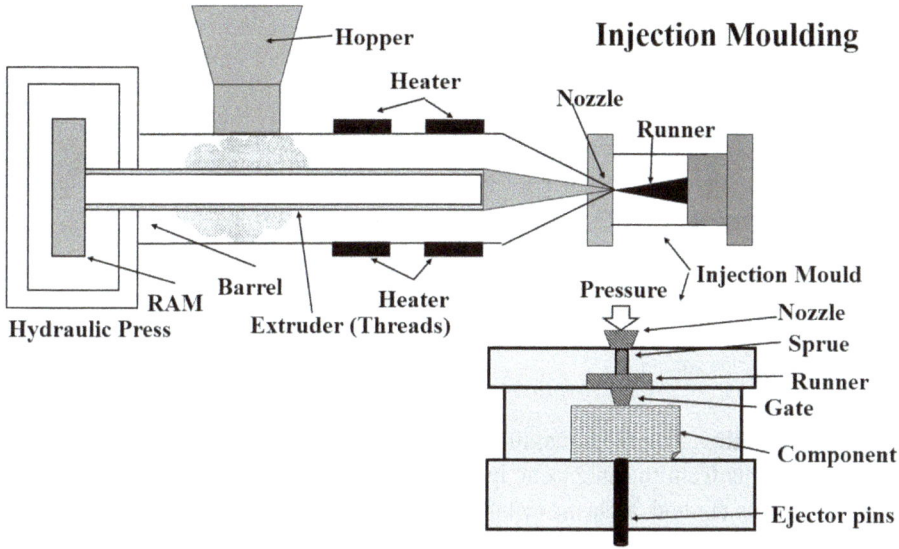

Figure 7.19(a): Injection molding.

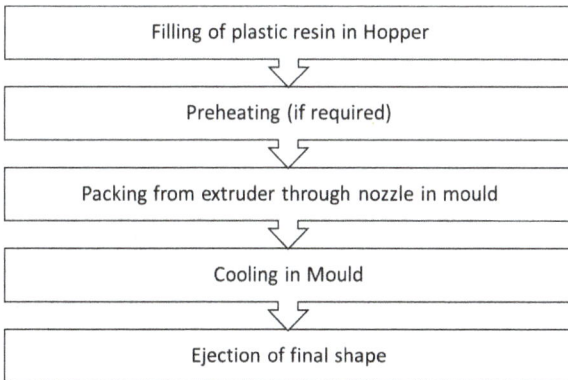

Figure 7.19(b): Injection molding steps.

However, thermosetting materials cannot be processed using this type of extrusion equipment, as they harden too quickly. Instead, a specialized extrusion machine for thermosets uses a ram instead of a screw to push the material through the die. The material is fed through a hopper, and repeated strokes of the ram force it into a long, tapered die with a heated zone. Curing is completed as the material reaches the forward end of the die. A schematic diagram of extrusion molding is shown in Figure 7.20.

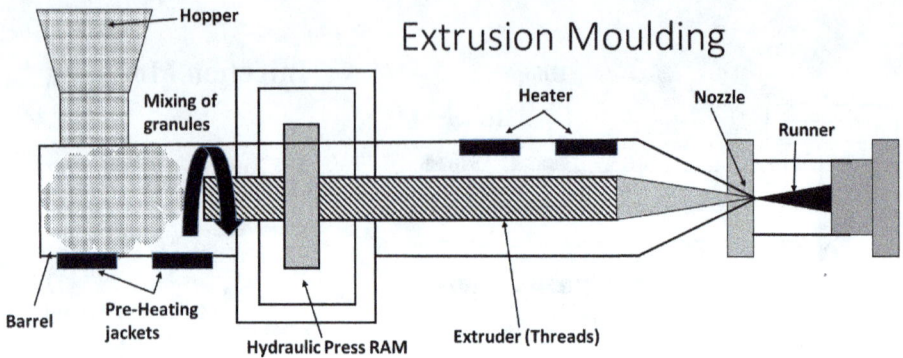

Extrusion Moulding

Figure 7.20: Extrusion process.

v. **Blow molding:** The blow molding process is used to create thin-walled, hollow components from thermoplastic materials such as polyethylene and polypropylene. In this method, a plastic cylinder (known as a parison) is extruded and positioned at the center of a two-part open mold, as illustrated in Figure 7.21(a). The material is then cut off, and the mold is closed. Compressed air is introduced into the hollow tube, causing it to expand and conform to the shape of the mold. Once the material has cooled, the mold is opened, and the finished component is removed. Common blow-molded products include bottles, containers, and floats. The steps of operation are represented in Figure 7.21(b).

7.8.4 Basic elements of a two-plate metal mold

A two-plate metal mold is a fundamental tool in injection molding, commonly used for manufacturing plastic parts. It consists of two main plates that come together to form the cavity where the molten material is injected and cooled. Here are the basic elements (see Figure 7.22) of a two-plate metal mold:

i. **Core plate:** The core plate contains the core of the mold, which shapes the interior of the part. It is often fixed to one side of the mold assembly.

ii. **Cavity plate:** The cavity plate contains the cavity, which shapes the exterior of the part. It is typically fixed to the opposite side of the core plate and aligns with it to form the complete mold cavity.

iii. **Mold base:** The mold base is the framework or housing that holds the core and cavity plates together. It provides the necessary support and alignment for the mold components.

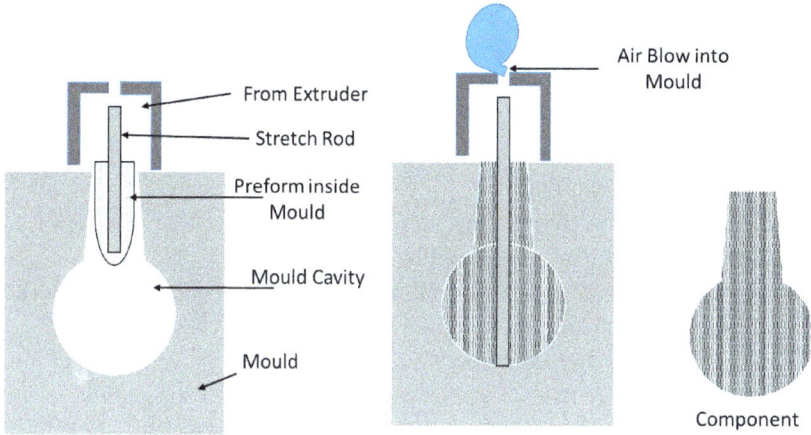

Blow Moulding

Figure 7.21(a): Blow molding process.

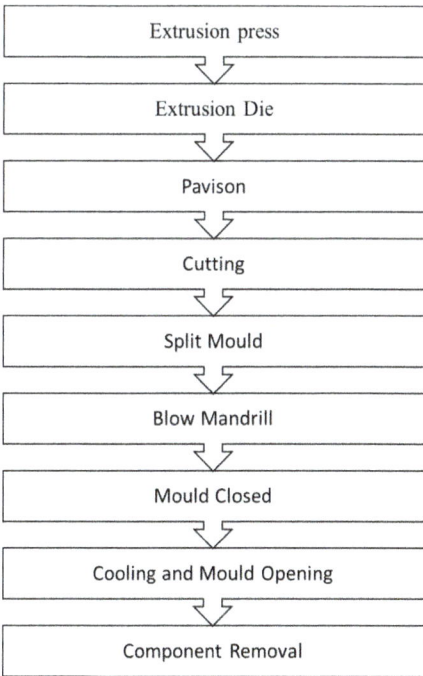

Figure 7.21(b): Blow molding process steps.

iv. **Ejection system:** This system ejects the molded part from the mold once it has cooled and solidified. Common ejection mechanisms include ejector pins, sleeves, and plates.

v. **Cooling channels:** These are passages within the mold that allow coolant (usually water) to flow through and regulate the temperature of the mold, ensuring that the plastic solidifies properly.

vi. **Sprue:** The sprue is the channel through which molten plastic is injected into the mold cavity from the injection molding machine.

vii. **Runner system:** The runner system channels the molten plastic from the sprue to the various cavities within the mold if there are multiple cavities.

viii. **Gates:** Gates are the entry points where molten plastic flows from the runner into the mold cavity. They are strategically placed to ensure proper filling of the cavity.

ix. **Vents:** Vents are small openings that allow trapped air and gases to escape from the mold cavity during the injection process, preventing defects in the molded part.

x. **Clamping mechanism:** This mechanism secures the mold plates together during the injection process and ensures that they remain aligned.

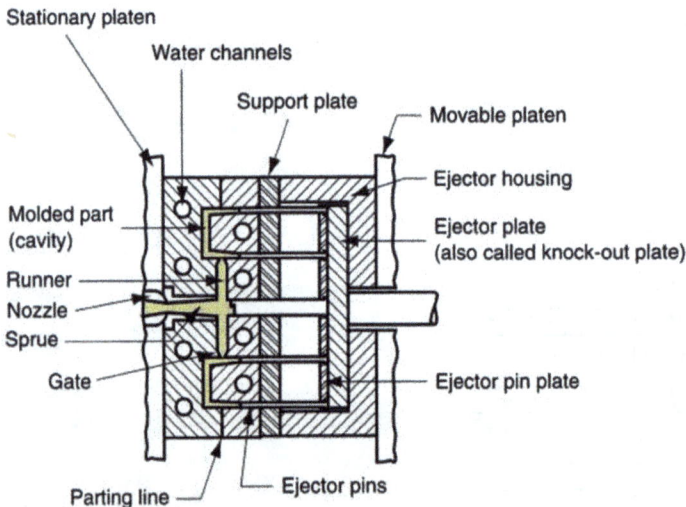

Figure 7.22: Basic elements of a two-plate metal mold.

7.9 Galvanizing and electroplating

i. **Galvanizing:** Galvanizing is a process of coating zinc onto iron or low-carbon steel. This process is carried out by immersing the work metal in a molten zinc bath. There are various ways to galvanize the zinc surface, and one of the most economical methods is the hot-dipping process.

a. **Galvanization by the hot-dipping process:** First of all, clean the surface of the workpiece and flux it by dipping the material in a solution of zinc chloride and hydrochloric acid. This will lead to protection from oxidation of the metal when it enters the molten bath. The molten metal bath is covered with a molten flux layer. Next, dip the metal in the zinc bath. The duration of immersion in the bath may depend on the mass of the article being coated. After that, allow it to cool and dry after withdrawing it from the zinc bath. The basic steps of the galvanizing process are shown in Figure 7.23.

Applications: Pipes, sheeting for roofs and walls of buildings, canes, outdoor hardware, etc.

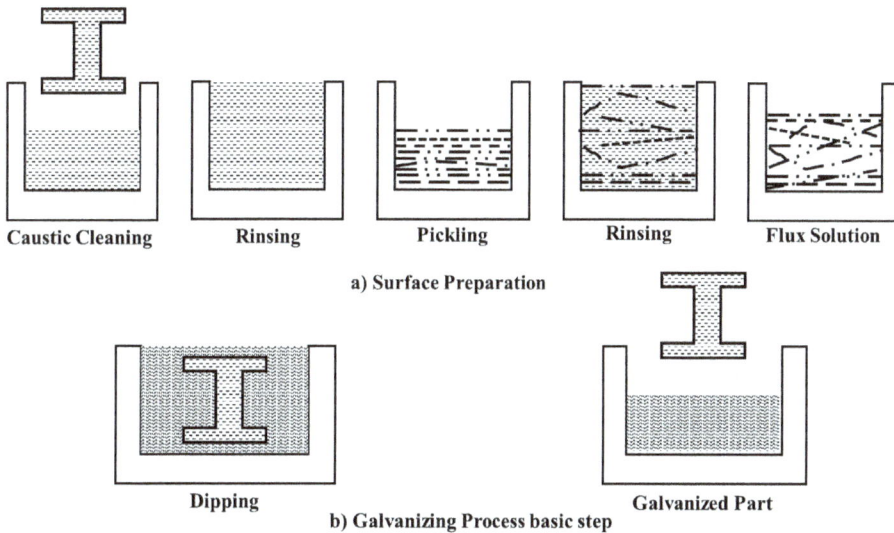

a) Surface Preparation

b) Galvanizing Process basic step

Figure 7.23: Galvanizing process: basic steps.

ii. **Electroplating:** It is a process of plating a metal with a different metal. In this process, the workpiece and plating material are both suspended in a bath containing a water-based electrolytic solution (CuSO4). Here, the plating material acts as the anode, and the workpiece acts as the cathode. These electrodes (cathode and anode) are connected, and the metal ions from the anode, under potential from the external source of electricity, combine with the ions in the electrolytic bath and are deposited on the cathode. Electroplating is performed to make surfaces corrosion-resistant, wear-resistant, and to improve their appearance. Coating up to 0.45 mm thickness can be achieved by this method. Common coating materials include chromium, zinc, copper, cadmium, and tin. A schematic diagram of the electroplating process is shown in Figure 7.24.

Application: Generally used for developing wear and corrosion resistance of tools, hydraulic shafts, aircraft, and diesel engine cylinder liners, etc.

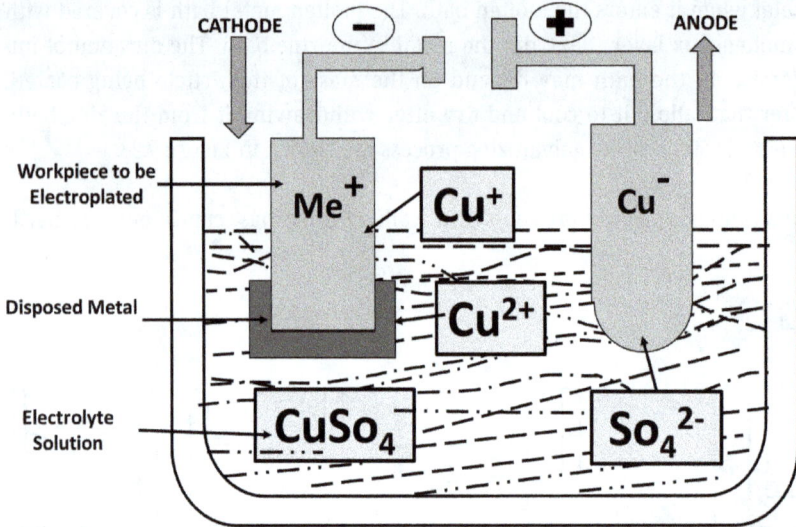

Figure 7.24: Electroplating principle.

7.10 Modern trends in manufacturing

Achieving success and gaining a competitive edge in the manufacturing industry require a well-defined and continuously evolving strategy. This strategy should integrate people, technology, focus, and direction while maintaining a long-term vision for planned investment and implementation.

A key trend in modern manufacturing is developing the right strategy and ensuring its consistent execution. In supply chain management, success depends on an iterative evaluation of cost-benefit trade-offs in operational components. Clearly defined supply chain objectives must be met, yet many companies lack a comprehensive supply chain management (SCM) strategy. To remain competitive, organizations should adopt an SCM approach that prioritizes supplier and vendor management while optimizing costs and efficiency. A consistent supply chain strategy is essential for effective interactions with suppliers, distributors, and customers. As markets become more interconnected, collaboration is critical. Well-executed supply chain management enhances value not just for individual organizations but for the entire supply chain.

In marketing, building trust within communities is crucial for success. With increasing web usage and expanding digital applications, seamless access to online platforms is becoming a necessity. As wireless device capabilities advance, many tasks currently performed on office computers will shift to mobile devices, fostering a globally connected community through platforms like Facebook.

Looking ahead, companies that leverage the interconnectedness of customers, partners, suppliers, and employees will emerge as industry leaders. Manufacturers must focus on continuously improving their solutions and software to align with customer needs. The true measure of success lies in strategies that strengthen customer relationships, drive revenue growth, and enhance profitability. Businesses that bring themselves closer to their customers while achieving these goals will secure long-term success in an evolving marketplace.

Keys to success and competitive advantage in manufacturing industries

i. **Focused manufacturing strategy**
   - Success in manufacturing requires a well-defined strategy that is continuously updated.
   - A balanced mix of people, technology, focus, and direction is essential.
   - Long-term planning is necessary for strategic investment and effective implementation.

ii. **Evolving manufacturing trends**
   - Developing the right manufacturing strategy and ensuring its adherence are crucial.
   - Supply chain success relies on the iterative evaluation of cost-benefit trade-offs in operational components.
   - Clear supply chain objectives must be established and met.

iii. **Effective supply chain management**
   - Many companies lack a comprehensive supply chain management (SCM) strategy.
   - Organizations should adopt an approach focused on supplier and vendor management.
   - The goal is to minimize costs while maximizing efficiency and competitiveness.
   - A consistent supply chain approach is necessary for effective interaction with suppliers, distributors, and customers.
   - As competition intensifies and markets become more interconnected, collaboration is the key.
   - Well-executed SCM adds value not only to the organization but also to the entire supply chain.

iv. **Leveraging digital connectivity for market success**
   - Successful marketers build trust within their communities.
   - Increased web usage fosters continuous connectivity and content sharing.
   - The rise of wireless access devices will enable tasks currently performed on office computers.
   - Social networks like Facebook enhance global connectivity.

v. **Future business success in 2020 and beyond**
   – Companies that leverage the interconnection of customers, partners, suppliers, and employees will thrive.
   – Manufacturers must continuously improve solutions and software to align with customer needs.
   – Strategies that strengthen customer relationships, drive top-line growth, and improve profitability will lead to long-term success.

7.11 Automation

The meaning of automation is derived from the term "automatic," a process which is carried out automatically without human effort is termed as automation. Automation utilizes mechanical advantages, computers, electronics, and electrical amplification to advance the production process. This automatic system helps provide guidance for operations, inspection of processes, and maintenance of the system with less human effort.

Generally, automation is defined as the automatic process that uses mechanical, electrical, electronic, and computer-based systems to operate and control production in an industry. In this system, the process is carried out fully or partially according to previously set programs, and there is no involvement of human activity during this automatic processing. Automation is used in various fields within production systems; some of them are listed below:

i.   To provide easy material handling within the system.
ii. It provides an automatic machine tool to process or cut the workpiece as per the requirement.
iii. Automation provides industrial robots, which help increase the productivity of the plant.
iv. In the quality control section, it provides an automatic inspection system that reduces inspection time and human error.
v.   This system broadly utilizes computers for planning, data collection, and designing.
vi. Automation provides firefighting robots, which reduce hazards in the workplace.
vii. Intelligent robots are capable of making decisions by themselves. They are useful in many areas of production.

7.11.1 Computer-aided design/computer-aided manufacturing (CAD/CAM)

Computer-aided design and computer-aided manufacturing are systems in which computer-based tools are used for designing, drafting, manufacturing, and controlling processes in a production system.

In CAD, it involves activities that are used to create engineering designs with the help of computers. Computers use computer graphics (interactive computer graphics) to create engineering designs. On the other hand, CAM uses industrial robots and CNCs to manufacture a product. CNCs and robots are programmed machines that follow the commands previously given to them. The movement of tools is controlled by computers through programming.

i. **NC machine:** Numerical control (NC) is a method used to control the production process. In an NC system, coded numerical instructions are inserted in to the machine tool to control the production processes. NC machines are automatic tools that are operated by programmed instructions encoded on a tape (storage medium). This eliminates the need for manual control of machines.
    –   **Advantages of NC machines over conventional machine control**
        a) Ease of complex part production.
        b) During difficult part production, special jigs and fixtures were needed; however, in the NC system, there is no need for special jigs and fixtures.
        c) It reduces set-up time to locate the part.
        d) Flexibility is increased.
        e) Due to the high accuracy of the product, inspection time is reduced.

ii. **Computer numerical control:** In computer numerical control (CNC), a program is built into the system to perform functions as needed. It consists of a tape reader for program entry, a computer for software functions, an MCU (Machine Control Unit), and a machine tool. The program tape is inserted into the tape reader, and the computer processes the instructions given on the tape and transfers them to the MCU. The MCU contains electronic circuits that read the program and interpret it into the mechanical movement of the machine tool. The complete process layout for the CNC system is shown in Figure 7.25.

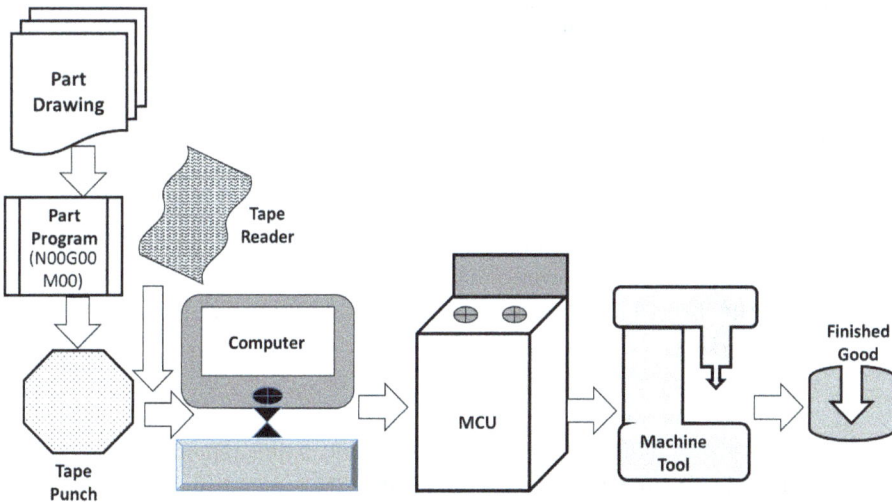

Figure 7.25: Complete process of the CNC system.

- **Advantages of CNC**
  a) More flexible as modifications are made in the program.
  b) It reduces data-reading errors.
  c) Program can be easily modified in the data stored in computer memory.
  d) Can detect machine malfunctions before the product is completely made.
  e) New options can be added to the system easily and at a low cost.
  f) Easy conversion of units. It can convert from the British to the International System of Units and vice-versa.

iii. **Direct numerical control (DNC):** It can be defined as a manufacturing system in which a number of NC machines are controlled by a single computer. The instructions are directly transferred to the machines from the main computer. There is a communication line that helps transfer data/instructions to the machine tool. When instructions are being transferred, both the computer and the machine tool send data to each other simultaneously. The storage capacity of the computer memory is very high, allowing it to control more than 120 machines at a time.

- **Advantages of DNC**
  a) It controls more than one machine with a single computer.
  b) Enhanced computational capability for operations such as circular interpolation, linear interpolation, tapering, etc.
  c) There is no need for tape and a tape reader. Here, the tape and tape reader are replaced by communication cables.
  d) More reliable.
  e) There is no need for a hardwired control unit in DNC.

iv. **Flexible manufacturing system (FMS):** When a number of CNCs are connected with automatic material handling equipment to produce a product, and the whole system is controlled by a central computer (master computer). During production, if this system is allowed to change the design of the product randomly, this type of manufacturing is called a flexible manufacturing system. This is an automated manufacturing system where sufficient flexibility is provided to meet a wide range of customer demands. An FMS typically contains a set of CNC machine tools, conveyors, and automated guided vehicles (AGVs).

- **Advantages**
  a) Improved capital utilization, as the changes from one part to another are faster and achieved at a low cost
  b) The direct labor cost is reduced due to a reduction in the number of workers
  c) Inventory is reduced
  d) Higher productivity
  e) Reduced indirect costs due to less rework, errors, repairs, etc.

v. **Computer integrated manufacturing (CIM):** CIM is an advanced form of CAD/CAM. In a CIM system, the computer is widely used to design the product, control the operation, plan the production, and perform business-related tasks. CIM includes computer-aided process planning, CAD/CAM, and FMS to perform the entire range of tasks related to the production system. The block diagram in Figure 7.26 shown below explains the various tasks performed under CIM.

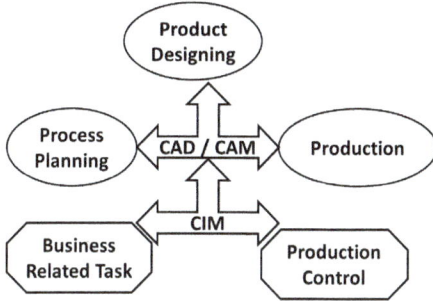

Figure 7.26: Functioning of computer-integrated manufacturing.

7.12 Industrial safety and health

Safety is the art of minimizing or eliminating the probability of a hazard that may cause an accident. Accidents are the invasion of the unprepared by the unexpected.

Industrial accidents are mainly due to human failure – somewhere in the chain of circumstances leading up to injury, we find a human factor. The main question now is, "Who failed?" The prevailing attitude to this day, even among workers, is to blame the injured person himself. At the shop supervisory level, there is little regard for safety rules, and at the management level, unfortunately, one can still find ignorance of legal and moral responsibilities. There must be recognition of the rightful place of accident prevention in the overall management function. The challenge of accident prevention is to change this attitude by providing regular safety training and making people safety-conscious at all levels, not only for their own personal safety but also for the safety of others.

An attitude needs to be developed that causes one to routinely consider the possibility of accidents in each and every situation and activity in which they are involved, and to take suitable preventive measures and actions to eliminate the possibility of accidents.

7.12.1 Safety concept and definition

There is always a safe and logical method of performing a task. Unsafe methods cause the majority of accidents. Method improvements by work-study engineers can also reduce the number of industrial accidents by 75%.

Safety should be built into the jobs. While planning for safe methods for performing a job, works managers, production managers, work-study engineers, and safety officers must ensure that all aspects of safety have been considered and addressed. Safety considerations should begin at the factory planning level, starting from the architect's board, and extend to the control and operation of machinery, as well as the consideration of illumination and noise.

7.12.2 Causes and common sources of accidents

With the foregoing introduction, let us now consider the human element and its interactivity with events that lead to accidents. There are always antecedents (unsafe conditions, unsafe events) that lead to unsafe incidents, which cause accidents or close calls as their consequences.

The interaction of the human element with the event that leads to accidents can be explained by means of a simple example. A woodsman chops down a tree, which can crash down on a car. This is an accident. If there is no car in the vicinity, it becomes an unsafe act that results in an unsafe environmental event.

Now, the driver's (human) interaction with the hazard becomes an unsafe incident, and it may cause an accident, a near accident, or accident potential depending upon the alertness of the driver:

a) The driver hit the tree – accident.
b) The driver applied the brakes late, almost (but not quite) hitting a tree – near accident.
c) The driver sees the fallen tree well in advance and applies the brakes to stop without any apparent risk of accident potential.

7.12.3 Industrial accidents

Accidents are the intrusion of the unprepared by the unexpected. An accident is an undesirable, unexpected event that happens to the unprepared. An accident disrupts the normal functioning of a person, causing injury or near-injury. Industrial accidents are mainly due to human failure. Somewhere in the chain of circumstances leading to the injury, we find the human factor.

7.12.4 Types of accidents

Accidents may be classified as follows:
i.   Based on the length of recovery from injury
ii. Cause of events
iii. Nature of events

iv.   Nature of injury
v.   Accidents on construction work

i.   **Based on the length of recovery from injury:**
  a.   **First aid cases:** the workers are given first aid at the factory hospital and resume work after treatment. There is no compensation.
  b.   **Home case accidents:** In this case, after receiving first aid, the worker is allowed to go home but is expected to resume duty the next day. There is no compensation.
  c.   **Lost time accident:** For these accidents, the lost time is more than one day, the injured person is hospitalized, may have temporary or permanent disablement, and the cause of the accident is investigated.

ii. **Cause of events:**
  a.   **Machine accidents:** catching of body parts or clothing in machines, catching of tools in machines, or catching of flying objects or particles in machines.
  b.   **Non-machine accidents:** These accidents are common but less serious, e.g., falling objects on the floor, pushes, bumps, or other persons' objects.

iii. **According to nature of events:**
  a.   **Traffic accidents:** Collision with pedestrians, collision with cars, or collision with motor vehicles.
  b.   **Passenger accidents:** These may include boarding moving vehicles, being caught or struck by doors, or slipping and stumbling.

iv. **According to the damage caused:** Damage could be to property, materials, or buildings. Some examples are damage to stored materials, loss of containers/contents, and damage to hand trucks, trolleys, conveyors, cranes, or machines.

v.   **According to the nature of the injury:** This classification is as follows:
  a.   **Fatal accident:** One or more persons are killed.
  b.   **Permanent disability:** Due to an accident, workers lose earning capacity. Compensation is paid depending on the extent of disablement (partial/total).
  c.   **Temporary disablement:** These accidents are less serious than the previous category. The worker becomes temporarily unfit (e.g., a fracture in the arm) and temporarily loses their earning capacity. Compensation is provided according to the Workmen's Compensation Act.

vi. **Accidents in construction work:** Some examples are as follows: falling of a person, slipping resulting in body strain, being struck by a flying object, accidents due to traffic, burns or fire accidents, electric shocks, explosions, etc.
  –   **Causes of accidents:** There are always some causes of accidents. To avoid accidents, these causes must be studied in detail. These causes may be divided into the following categories.
    a)   Physical causes due to unsafe work conditions
    b)   Accident due to workers' defective physiology
    c)   Accident due to the mental state of the worker

7.12.5 Methods to enhance safety in industry

Steps to enhance safety in industry

i.   **Personal protective equipment:** Protective clothing should be worn wherever necessary, such as gloves, which protect workers' hands from cuts, splinters, and chemicals. Aprons protect the trunk of the body, while metallic foot and leg guards, as well as protective helmets, save workers from falling objects. Women workers use hair pins and nets to ensure their hair does not get caught in gears or spinning shafts. During welding, goggles and face shields are used. If the noise is unbearable, earplugs are used. Safety shoes protect legs from prickling substances and exposure to heat.

ii. **Machine guarding:** Machine running parts must be properly guarded. These guards should provide full protection and should not be removable. Permanent rotating parts should not be operated beyond the recommended speed values. In the case of presses and shears, hazards can be reduced by using automatic feeds.

iii. **Protection against radiation hazards:** Radiation can be categorized into two types: heat radiation and radiation from radioactive particles. Both types pose health and safety risks, causing excessive perspiration, exertion, and skin irritation. Plants and equipment are designed to minimize radiation exposure. Proper ventilation, air conditioning, and protective clothing help mitigate these issues.

iv. **Noise control:** Excessive noise beyond 100 dB is dangerous and harmful. Office work produces 50 dB, and conversation generates 40 dB of noise, which is considered normal. A pneumatic chisel produces 90 dB of noise.

v.   **Fatigue:** fatigue also causes accidents. There are more accidents at the end of working hours. Proper rest pauses can help eliminate fatigue during work.

vi. **Proper lighting:** Many accidents occur due to improper lighting. Light sources include natural light and artificial light (tubes and bulbs). When natural light becomes insufficient, artificial light is used.

7.12.6 First aid

i.   **Aims:** A first aid program aims to provide the most adequate possible treatment for all injuries that occur despite the best efforts to prevent them. The first-aid measures are taken prior to the arrival of medical personnel or to transport the patient to a medical facility. Every worker should know how to obtain medical assistance quickly and what actions to take until it becomes available. This requires:
     a) Fully competent first-aid attendants
     b) Fully equipped first-aid room
     c) Proper organization and record
     d) Employee co-operation

Occupational Safety and Health Administration **(OSHA)** standards require that, where medical personnel are not close to a worksite, at least one person trained in first-aid must be available.

ii. **Immediate action:** The actions taken during the first few minutes are extremely important. Well-intentioned but inept assistance may be worse than no assistance. The following injuries require prompt action:
   a) Asphyxiation or inhalation of toxic gases/liquids
   b) Electric shock
   c) Corrosive burns or cryogenic liquid splashes.
   d) Wound and cuts
   e) Thermal bums
   f) Bone fractures

In an accident in which an incapacitating injury has occurred, the injured person should be moved as little as possible, except under the following circumstances.
   a) **Asphyxiation or exposure to toxic gas:** The victim should be moved to fresh air if air or oxygen cannot be supplied where the victim is.
   b) **Electric shock:** If the person is still in contact with a live line, shut-off the power. If this is not possible, remove them or the conductor using a wooden or non-conducting pole.
   c) **Corrosive burns from cryogenic liquid splashes:** Wash off the corrosive or cryogenic substance. The injured person should be placed under an emergency shower or washed with water from a hose. Any clothing saturated with the harmful liquid should be removed, and the skin should be flushed with a large amount of water.
   d) **Fire:** The person should be removed to a safe area. The person giving assistance should be careful not to become a casualty. They should take prompt action while remaining calm. If the victim needs to be moved, it must be done carefully. Check for injuries to determine where care is required during movement (if time permits). The victim can be pulled to safety by one person using a blanket, tarpaulin, or a sheet of heavy plastic. Pulling should be along the body axis, not sideways.

Again, note that a person should not be moved when incapacitated unless there is imminent danger where the victim lies.

iii. **Aiding the victim**
   a) **Check the victim for breathing and bleeding:** Artificial respiration should be given immediately to any person whose breathing has stopped. Mouth-to-mouth or other means of resuscitation combined with heart massage should be maintained until breathing is restarted or medical help arrives. Only inhalators that induce lung action are a suitable substitute for artificial respiration. Any uncontrolled bleeding should be stopped by compressing the

wound using a dressing from a first-aid box, a clean handkerchief, towel, or shirt applied with firm pressure.

b) **Medical assistance:** Medical assistance should be requested as soon as possible.

c) **Shock:** Any person involved in a serious accident should be considered to be in shock and treated accordingly. It has frequently happened that a person involved in an accident has appeared to be uninjured, walked away, and later collapsed and died. A person who has been in an accident should be kept quiet and warm, protected from chills with a coat, blanket, etc. If there is no head wound or broken leg, the head should be lowered and the legs should be raised slightly to help increase blood flow to the brain. When there is a head wound, the head should be slightly raised or kept in level with the body.

d) **Unconsciousness:** If the injured person is unconscious but breathing, they should be placed face down with their head turned to one side to prevent choking in case they should throw up. Once in position, the victim should not be moved again until medical help arrives. Do not attempt to give a drink to an unconscious person.

e) **Fracture:** A person with a fracture must be treated carefully to prevent the injury from worsening and to reduce the risk of shock. A fracture is suspected if:
   I.   The part has abnormal shape
   II. There is an inability to move the part and pain during movement.
   III. There is swelling and a change in the color of the skin.

f) **Burns:** Immersing a burnt limb in cold water or applying cold water to the burned area reduces pain and shock. Medical assistance should be obtained. Do not apply oil, grease, or butter to the burnt area. Do not touch the burnt part.

Questions

Long answer-type questions

1. Why is galvanizing preferred over paint coatings for steel structures exposed to marine environments? Discuss with examples. **(GATE 2018)**

2. Analyze the role of preventive maintenance in ensuring safety in an automated plant. **(GATE 2020)**

3. Compare the applications of CNC and DNC in automated machining processes. Discuss their advantages and limitations. **(IES 2020)**

4. Describe blow molding and its types. How does the Parison design affect the final product's quality? **(IES 2020)**

5. Explain the significance of industrial hygiene and the steps involved in ensuring a safe workplace. **(IES 2021)**

6. Describe the advantages of flexible manufacturing systems (FMS) in improving automation and productivity in industries. **(GATE 2021)**

7. Explain the factors influencing the selection of wood for furniture making. What are the advantages of using hardwood over softwood in specific applications? **(GATE 2021)**

8. Describe methods for mitigating risks in a hazardous manufacturing environment. **(GATE 2022)**

9. Explain the injection molding process. Discuss how process parameters, such as injection pressure and cooling time, influence the quality of the final plastic product. **(GATE 2022)**
10. Explain the different types of plant layouts (process layout, product layout, and cellular layout). Discuss the advantages and limitations of each type with respect to the manufacturing process.
(GATE 2022)

Short answer-type questions

1. Briefly explain fixed-position layout and provide an example of an industry where it is commonly used. **(GATE 2019)**
2. Explain the concept of a product layout and its advantages in mass production. **(IES 2020)**
3. Define productivity. How can productivity be improved in a manufacturing process? **(GATE 2021)**
4. Describe the process of rotational molding. What are its advantages in making hollow plastic products? **(IES 2021)**
5. Explain the difference between metals and non-metals in terms of conductivity and malleability. Mention one non-metallic material and its industrial use. **(GATE 2021)**
6. How does manufacturing contribute to economic development? Mention two key benefits. **(GATE 2022)**
7. What is the difference between batch production and continuous production? Mention one advantage of each type. **(GATE 2022)**
8. What is job production? How does it differ from mass production? **(IES 2022)**
9. What is blow molding? Explain its application in the manufacturing of plastic bottles. **(IES 2022)**
10. Define polymers and ceramics. Discuss one property that distinguishes polymers from ceramics in terms of their mechanical behavior. **(IES 2022)**

Objective questions

1. Which material property is most significant for selecting materials in sustainable manufacturing processes? **(GATE 2022)**
    A. Thermal conductivity   B. Recyclability   C. Tensile strength   D. Electrical resistivity
2. Socio-economic development in manufacturing is primarily driven by: **(IES 2020)**
    A. Reduced labor requirements
    B. Automation and the use of advanced materials
    C. Shorter product lifecycle
    D. Higher consumer spending
3. Which of the following socio-economic factors is most critical in the adoption of composite materials in manufacturing? **(NET 2021)**
    A. Availability of raw materials
    B. Cost of production
    C. Ease of machining
    D. Esthetic appeal
4. In which type of plant layout are machines and equipment grouped according to the sequence of operations? **(GATE 2021)**
    A. Process layout   B. Product layout   C. Fixed-position layout   D. Cellular layout
5. The shipbuilding industry primarily follows which type of plant layout? **(IES 2020)**
    A. Product layout   B. Process layout   C. Fixed-position layout   D. Combination layout
6. Which type of plant layout is most suitable for manufacturing products with varying production processes? **(NET 2022)**
    A. Product layout   B. Process layout   C. Fixed-position layout   D. Cellular layout
7. Which type of production is characterized by high volume and low variety of products? **(GATE 2022)**
    A. Job production   B. Batch production   C. Mass production   D. Project production

8. In which type of production system is the product manufactured according to customer specifications
    and delivered as a single unit?   **(IES 2020)**
    A. Batch production
    B. Job production
    C. Continuous production
    D. Mass production
9. Which of the following is a characteristic of batch production?   **(NET 2021)**
    A. Flexible resource allocation
    B. Continuous flow of materials
    C. High product standardization
    D. One-off production of items
10. Which type of wood is primarily used for making furniture due to its durability and esthetic appeal?
                                                                             (GATE 2021)

    A. Teak   B. Pine   C. Balsa   D. Bamboo
11. Softwoods are generally obtained from which type of trees?   **(IES 2020)**
    A. Deciduous trees
    B. Evergreen coniferous trees
    C. Fruit-bearing trees
    D. Tropical hardwood trees
12. Which of the following is a characteristic of hardwoods compared to softwoods?   **(NET 2022)**
    A. Faster growth rate
    B. Higher density and durability
    C. More uniform grain structure
    D. Limited applications in furniture
13. Which of the following is a primary advantage of using ceramics in high-temperature applications?
                                                                             (GATE 2022)

    A. High ductility
    B. Excellent thermal insulation
    C. Superior electrical conductivity
    D. Easy machinability
14. Thermoplastics are distinct from thermosetting plastics because they: **(IES 2020)**
    A. Can be reshaped upon reheating
    B. Has higher thermal resistance
    C. Are more brittle after molding
    D. Require a curing process
15. Which of the following best describes a composite material?   **(NET 2021)**
    A. A homogeneous material with uniform properties
    B. A combination of two or more materials with distinct physical and chemical properties
    C. A material made from a single polymer matrix
    D. A metal alloy with improved conductivity

Answers
1(B)   2(B)   3(B)   4(B)   5(C)   6(B)   7(C)   8(B)   9(A)   10(A)   11(B)   12(B)   13(B)   14(A)   15(B)

Bibliography

A. B. Badiru, V. V. Valencia, D. Liu. "Additive Manufacturing Handbook – Product Development for the Defense Industry", CRC Press, 2017.

A. S. Darmawan, B. W. Febriantoko, A. D. Anggono, T. W. B. Riyadi, A. Hamid. Effect of Thickness Reduction on Cold Rolling Process to Microstructure and Brass Hardness, In: "MATEC Web of Conferences", 2018.

A. Kumar, P. Kumar, A. K. Srivastava, L. Saharan. "Manufacturing Strategies and Systems – Technologies, Processes, and Machine Tools", CRC Press, 2025.

A. Livesey, A. Robinson. "The Repair of Vehicle Bodies", Routledge, 2018.

A. Villanueva-Merino, S. Urra-Uriarte, J. Izkara, S. Campos-Cordobes, A. Aranguren, P. Molina-Costa. "Leveraging Local Digital Twins for planning age-friendly urban environments", Cities, 2024, 155, 18. https://doi.org/10.1016/j.cities.2024.105458.

B. Black. "Workshop Processes, Practices and Materials", Routledge, 2019.

C. G. Hussain, M. Qadeer, R. Keçili, C. M. Hussain. Additive manufacturing in the next world, In: "Medical Additive Manufacturing Concepts and Fundamentals", Elsevier BV, 2024, 299–362. https://doi.org/10.1016/B978-0-323-95383-2.00007-X.

C. W. Gellings. "Saving Energy and Reducing CO_2 Emissions with Electricity", Routledge, 2020.

D. Dhinakaran, N. Jagadish Kumar, A. Raja Brundha, N. Subiksha, T. Karpagam. "Chapter 2 Precision Maintenance with PARM and Augmented Reality for Asset Optimization", In: Navigating the Augmented and Virtual Frontiers in Engineering, IGI Global, 2024, 21–48. DOI: 10.4018/979-8-3693-5613-5.ch002.

D. Hui, R. D. Goodridge, C. A. Scotchford, D. M. Grant. "Laser sintering of nano-hydroxyapatite-coated polyamide 12 powders", Additive Manufacturing, 2018, 22, 560–570. https://doi.org/10.1016/j.addma.2018.05.045.

D. K. Singh. "Fundamentals of Manufacturing Engineering", Springer Science and Business Media LLC, 2024.

D. Kim, M. Ramulu. "Drilling process optimization for graphite/bismaleimide–titanium alloy stacks", Composite Structures, 2004, 63(1), 101–114. https://doi.org/10.1016/S0263-8223(03)00137-5.

D. Doran, B. Cather. "Construction Materials Reference Book", Routledge, 2013.

D. F. Tver, R. W. Bolz. "Encyclopedic Dictionary of Industrial Technology", Springer Nature, 1984.

D. Pin. Chapter 215 Anti-seepage for Groundwater Pollution, Springer Science and Business Media LLC, 2024 In: "Proceedings of the 9th International Conference on Water Resource and Environment (WRE 2023) Series: Lecture Notes in Civil Engineering", Vol. 468, DOI: 10.1007/978-981-97-0948-9.

D. Atalie, Z.-S. Guo, D. Berihun, M. Tadesse, M. Peng-Cheng. "Role of Additive Manufacturing in Defense Technologies: Emerging Trends and Future Scope", In: Additive Manufacturing Materials and Technologies, Elsevier BV, 2024, 501–521. https://doi.org/10.1016/B978-0-443-18462-8.00020-9.

E. Sycheva, P. Shpak. "Chapter 3 Practical Application of the Concept of Digital Twins in the Aviation Sector", In: Digital Transformation in Industry. Lecture Notes in Information Systems and Organisation, Springer Science and Business Media LLC, 2022 vol. 54. https://doi.org/10.1007/978-3-030-94617-3_3.

G. K. Awari, V. S. Kumbhar, R. B. Tirpude, S. W. Rajurkar. "Automotive Manufacturing Processes – A Case Study Approach", CRC Press, 2023.

F. B. Gary. "Nontraditional Manufacturing Processes", CRC Press, 2017.

G. Boothroyd, W. A. Knight. "Fundamentals of Machining and Machine Tools", CRC Press, 2019.

G. Chryssolouris. "Chapter 2 Overview of Manufacturing Processes", In: Manufacturing Systems: Theory and Practice, Springer Science and Business Media LLC, 1992.

S. S. Gill. "An adaptive neuro-fuzzy inference system modeling for material removal rate instationary ultrasonic drilling of sillimanite ceramic", Expert Systems with Applications, 201008.

https://doi.org/10.1515/9783112205891-008

"Handbook of Electronics Packaging Design and Engineering", 1990.

"Handbook of Manufacturing Engineering and Technology", 2015.

H. Abdel-Gawad El-Hofy. "Fundamentals of Machining Processes – Conventional and Nonconventional Processes, Second Edition", CRC Press, 2019.

H. A. Youssef, H. A. El-Hofy, M. H. Ahmed. "Manufacturing Technology – Materials, Processes, and Equipment", CRC Press, 2019.

H. Youssef, H. El-Hofy. "Non-traditional and Advanced Machining Technologies", CRCPress, 2020.

H. Youssef, H. El-Hofy. "Traditional Machining Technology", CRC Press, 2020.

H. Soleimanzadeh, B. Rolfe, M. Bodaghi, M. Jamalabadi, X. Zhang, A. Zolfagharian. "Sustainable robots 4D printing", Advanced Sustainable Systems, 2023, 7(9), 1–41.

I. Onaji, D. Tiwari, P. Soulatiantork, B. Song, A. Tiwari. "Digital twin in manufacturing: conceptual framework and case studies", International Journal of Computer Integrated Manufacturing, 2022, 35(8), 831–858.

J. Paulo Davim. "Innovative Development in Micromanufacturing Processes", CRC Press, 2023.

J. Vinoth Kumar, K. Radhakrishnan, L. SrimathiPriya. "Chapter 13 Shaping Tomorrow: The Impact of 3D Printing on Functional Material Fabrication", In: Breaking Boundaries: Pioneering Sustainable Solutions Through Materials and Technology, Springer Science and Business Media LLC, 2025.

J. R. Sanford, B. Comisky. The doughnut lens, In: "2008 IEEE Antennas and Propagation Society International Symposium", 2008.

K. A. Padmanabhan, S. Balasivanandha Prabu, R. R. Mulyukov, A. Nazarov, R. M. Imayev, S. Ghosh Chowdhury. "Superplasticity", Springer Science and Business Media LLC, 2018.

K. Sajan, P. Eberhard, S. K. Dwivedy. "Dynamic analysis of cold rolling process using the finite element method", Journal of Manufacturing Science and Engineering, 2015, 138(4), 041002-1–041002-10.

K. Ukoba, T.-C. Jen. "Thin Films, Atomic Layer Deposition, and 3D Printing – Demystifying the Concepts and Their Relevance in Industry 4.0", CRC Press, 2023.

L. Alting. "Manufacturing Engineering Processes", CRC Press, 2020.

L. T. Temane, J. T. Orasugh, S. S. Ray. "Polymer Additive Manufacturing: An Overview", Elsevier BV, 2024.

L. Jin, X. Zhai, K. Wang, K. Zhang, D. Wu, A. Nazir, J. Jiang, W.-H. Liao. "Big data, machine learning, and digital twin assisted additive manufacturing: A review", Materials & Design, 2024, 244, 1–53.

M. M. Farag. "Materials and Process Selection for Engineering Design", CRC Press, 2019.

M. Nuriev, T. Aygumov, R. Zaripova, S. Nikolaeva, G. Gumerova. RETRACTED: Evolving network systems through block chain innovation for smart agriculture IoT networks, In: "BIO Web of Conferences", 2024.

M. Schwartz. "Encyclopedia of Materials, Parts and Finishes", CRC Press, 2019.

M. Tooley. "Engineering GNVQ: Intermediate", Newnes, 2012.

M. Soori, B. Arezoo, R. Dastres. "Digital Twin for Smart Manufacturing, A Review", Sustainable Manufacturing and Service Economics, 2023, 2, 1–21.

M. A. Ali Khan, A. K. Sheikh, B. S. Al-Shaer. "Evolution of Metal Casting Technologies", Springer Science and Business Media LLC, 2017.

N. K. Jain, S. Pathak. "3.23 Electrochemical Processing and Surface Finish", Elsevier BV, 2017.

N. K. Jain, V. K. Jain, K. Deb. "Optimization of process parameters of mechanical type advanced machining processes using genetic algorithms", International Journal of Machine Tools and Manufacture, 2007, 47(6), 900–919.

N. K. Jain, K. Gupta. "Spark Erosion Machining – MEMS to Aerospace", CRC Press, 2020.

P. V. Mohanan. "Compendium of 3D Bioprinting Technology", CRC Press, 2025.

P. Bhambri, S. Kumar, P. K. Pareek, A. A. Elngar. "AI-Driven Digital Twin and Industry 4.0 – A Conceptual Framework with Applications", CRC Press, 2024.

P. J. Bártolo, A. J. Mateus, F. D. Conceição Batista, H. A. Almeida et al. "Virtual and Rapid Manufacturing – Advanced Research in Virtual and Rapid Prototyping", CRC Press, 2007.

P. Hughes, E. Ferrett. "Introduction to Health and Safety in Construction", Routledge, 2019.

R. A. Higgins. "Materials for Engineers and Technicians", Routledge, 2019.

R. Kumar, M. M. Mahapatra, A. Pradhan, A. Giri, C. Pandey. "Experimental and numerical study on the distribution of temperature field and residual stress in a multi-pass welded tube joint of Inconel 617 alloy", International Journal of Pressure Vessels and Piping, 2023, 206, 1–10.

R. Singh. "Arc Welding Processes Handbook", Wiley, 2022.

R. W. Messler. "Principles of Welding", Wiley, 1999.

R. Timings. "Engineering Fundamentals", Taylor & Francis Group, 2007.

R. Timings. "Fabrication and Welding Engineering", Routledge, 2008.

S. M. Sapuan, Y. Nukman, N. A. Abu Osman, R. A. Ilyas. "Composites in Biomedical Applications", CRC Press, 2020.

S. Prasad Jones Christydass, N. Nurhayati, S. Kannadhasan. "Hybrid and Advanced Technologies", CRC Press, 2025.

S. N. Bhavsar, A. Y. Joshi, K. M. Tamboli. "4th International Conference on Innovations in Automation and Mechatronics Engineering (ICIAME2018)", IntechOpen, 2018.

S. Krishnan, A. J. Anand, S. Sendhilkumar. "Handbook of Industrial and Business Applications with Digital Twins", Routledge, 2023, 408. ISBN 9781003465904. https://doi.org/10.1201/9781003465904.

S. B. Singh, P. Ranjan, A. V. Vakhrushev, A. K. Haghi. "Mechatronic Systems Design and Solid Materials – Methods and Practices", Apple Academic Press, 2021.

S. Chakraborty, S. Dey. "Design of an analytic-hierarchy-process-based expert system for non-traditional machining process selection", The International Journal of Advanced Manufacturing Technology, 2006, 31(5–6), 490–500.

S. D. El Wakil. "Processes and Design for Manufacturing", CRC Press, 2019.

T. H. Allegri. "Handling and Management of Hazardous Materials and Waste", Springer Science and Business Media LLC, 1986.

"Transforming Industry Using Digital Twin Technology", Springer Science and Business Media LLC, 2024.

V. K. Jain. "Advanced Machining Science", CRC Press, 2022.

W. A. J. Chapman. "Workshop Technology – Part: IAn Introductory Course", Routledge, 2019.

W. Bolton, R. A. Higgins. "Materials for Engineers and Technicians", Routledge, 2020.

Z. Huda. "Metal Forming Processes", Springer Science and Business Media LLC, 2024.

Z. X. Khoo, J. E. Mei Teoh, Y. Liu, C. K. Chua, S. Yang, J. An, K. F. Leong, W. Yee Yeong. "3D printing of smart materials: A review on recent progresses in 4D printing", Virtual and Physical Prototyping, 2015, 10(3), 103–122.

About the authors

Er. Mahmood Alam, an esteemed Assistant Professor at Integral University, Lucknow, has built a remarkable 16-year career in academia, establishing himself as a prominent figure in the field of engineering. He was awarded a Silver Medal for his academic excellence during his postgraduate studies in Mechanical Engineering, specializing in Production and Industrial Engineering. Currently, Er. Mahmood is pursuing Doctoral studies at the prestigious *Indian Institute of Technology (ISM) Dhanbad*, focusing on the innovative field of microchannel heat sink design. His cutting-edge research is contributing significantly to advancements in this specialized domain. In addition to his research and teaching roles, Er. Mahmood is an active Member of ISHRAE (Indian Society of Heating, Refrigerating, and Air Conditioning), reflecting his dedication to professional development and industry engagement. His teaching expertise covers a wide array of subjects, including Manufacturing Science, Advanced Welding Technology, Polymer Science and Technology, Refrigeration and Air Conditioning, and HVAC (Heating, Ventilation, and Air Conditioning). Er. Mahmood's contributions to education have been recognized with the prestigious Excellence Award in Education by AMP, a notable NGO, underscoring his unwavering commitment to advancing knowledge. He has published numerous research papers and supervised various projects, further strengthening his impressive academic profile.

Dr. Abhishek Dwivedi, (MIE), Assistant Professor, Department of Mechanical Engineering, Integral University, Lucknow is having around 13+ years of educating and 1 year of Industrial experience. Dr Dwivedi is Diploma and Post Diploma in Plastic and Mould Technology, from CIPET. Lucknow (Govt. of India), B.Tech.-Mechanical Engineering, Integral University, Lucknow, M. Tech in Computer Integrated Manufacturing, from MMMEC., Gorakhpur, (UP State Govt.) under AKTU., Lucknow, and PhD. In Mechanical Engineering (Corrosion Engineering) from Integral University, Lucknow. His aptitudes are tasks on Advance Mechanical Softwares and CNC Machines (CAD/CAM). He has been awarded in the competitions like CAD (IUL), MAT- Lab (IITK) and best oral paper award from MRU, Faridabad. He is won also been recognized as "New Faces in Mechanical Engineering" by Marvelous Records Book of India. He has presented /attended / published a significant number of Research Article in SCI/Scopus/UGC at National and International conferences/seminars/Journals/workshops and organized as well. He has authored various books; focusing on scientific and research including text books as Project Management- (India), Introduction to Six Sigma, – Methods, Approaches and Applications, (India), Line Balancing: Today and Tomorrow: A Problem of Manufacturing Industry (Germany). He is related with numerous Society/Organizations/ journals as Member/Reviewer.

Prof (Dr)Anshuman Srivastava, Professor, Indian Institute of Packaging, Lucknow, Govt of India. Prof (Dr) Anshuman Srivastava, obtained his Ph.D. in the field of Material Science and Engineering from Indian Institute of Technology (BHU) Varanasi, India. He is having long industrial and academics experience. He has published several research papers in journals like Composites Science and Technology, Journal of Power Sources, Material Research Bulletin, Journal of composites, International Journal of Energy Research, The European Physical Journal Plus, *Materials Science & Engineering B*, Journal of Solid State Chemistry and many others. He is a member of advisory, organizing and technical committees in many national and international conferences and has successfully organized several international and national conferences / workshops and FDPs too. He has also delivered invited talks/expert lectures in several conferences and FDPs. He is member of Editorial Board and Reviewer panel of various international/national journals. Presently he is supervising UG/PG and doctorial theses in mechanical and material science field. He is life member of various prestigious professional bodies like Electron Microscopy Society of India, International Association of Engineers (IAENG), IAENG Society of Industrial Engineering, IAENG Society of Mechanical Engineering, Material Research Society of India, Indian Society for Technical Education, Asian Polymer Association, Indian Society for Advancement in Material Processing and Engineering, Senior Member Hong Kong Society of Mechanical Engineers SM No 20170612001, Hong Kong Chemical, Biological & Environmental Engineering Society. His research work

https://doi.org/10.1515/9783112205891-009

is focused on Physical, Dielectric and Mechanical properties of PVDF/CCTO and modified CCTO composites, Synthesis of high dielectric ceramics. Natural fibers reinforced Polymer composites, Bio composites, Metal matrix composites, Ceramics matrix composites.

Professor Ajay Kumar Mishra MPhil, PhD, CSci, FRSC is currently working as Professor in the Department of Chemistry, University of The Western Cape, Cape town, South Africa. He is "B" rated researcher recognized by National Research Foundation, South Africa. Prof Mishra has published over 450 publications in peer-reviewed international journals/conferences, books and book chapters. He has edited 40+ books and contributed over 80 book chapters in various peer reviewed edited books by established publishers. He has also delivered more than 150+ plenary/keynote/invited/guest lectures in various research institutions/universities/conferences/seminars/workshops. Prof Mishra has hosted many international visiting researchers and visited several universities globally. Prof Mishra's research has been cited more than 13300 times as per google scholar, with h-factor of 53. Recently, Prof Mishra have been named on a list of the top 2% of the most- cited scientists in various disciplines globally (2019-2024) by the data published by Stanford University, USA. Prof Mishra has been able to attract multimillion research grants from both internally and externally besides securing several research collaborations world-wide. Prof Mishra have attained considerable national and international recognition, as well as awards including "Fellow member" and "Chartered Scientist" by Royal Society of Chemistry, UK and Chancellor's Prize (Unisa) for excellent achievement in research. He was finalist for prestigious NSTF award for year 2016, 2023 and 2024. Prof. Mishra also serving as Associate Editor as well as member of the editorial board of many peer-reviewed international journals and books.

Index

https://doi.org/10.1515/9783112205891-010

www.ingramcontent.com/pod-product-compliance
Lightning Source LLC
Chambersburg PA
CBHW061339210326
41598CB00035B/5826